Tuffs—Their Properties, Uses, Hydrology, and Resources

Edited by

Grant Heiken
331 Windantide Place
Freeland, Washington 98249-9683

THE
GEOLOGICAL
SOCIETY
OF AMERICA

Special Paper 408

3300 Penrose Place, P.O. Box 9140 ▪ Boulder, Colorado 80301-9140 USA

2006

Published by The Geological Society of America, Inc.
3300 Penrose Place, P.O. Box 9140, Boulder, Colorado 80301-9140, USA
www.geosociety.org

Printed in U.S.A.

GSA Books Science Editor: Abhijit Basu

Library of Congress Cataloging-in-Publication Data

Tuffs : their poperties, uses, hydrology, and resources / edited by Grant Heiken.
        p. cm. — (Special paper ; 308)
    Includes bibliographical references.
    ISBN 13 978-0-8137-2408-9 (pbk.)
    ISBN 10 0-8137-2408-2 (pbk.)
        1. Volcanic ash, tuff, etc. I. Heiken, Grant. Special papers (Geological Society of America) ; 408

QE461.T84 2006
552.23--dc22

                                                                    2006048952

Cover: Remnants of the Claudian aqueduct, which has been constructed of tuff blocks quarried from the tuffs that underlie and surround the city of Rome, Italy. **Inset:** Campanian Ignimbrite, which forms the platform that underlies the city of Sorrento, Italy. **Background:** Wall at Villa Adriana, Tivoli, Italy. The small blocks of tuff have been cemented in a style called *opus vittatum*. **Back:** Images of Bandelier Tuff, Frijoles Canyon, New Mexico. This massive ignimbrite sequence was deposited during formation of the Toledo and Valles calderas, 1.6 and 2.1 Ma. Caves carved into the poorly consolidated base were used by people living on the Pajarito Plateau about 500 years ago.

10 9 8 7 6 5 4 3 2 1

Contents

Geological Society of America
Special Paper 408
2006

Tuffs—Their properties, uses, hydrology, and resources

edited by

Grant Heiken

331 Windantide Place, Freeland, Washington 98249-9683, USA

ABSTRACT

Tuffs have been a part of man's environment for thousands of years and underlie some of Earth's largest cities. There are 41 large industrial cities in 24 nations (including two megacities) that are underlain or partly underlain by tuffs. Rome, one of the world's most famous cities, has a history tied closely to the tuff deposits upon which the ancient city was built. Tuffs are products of explosive volcanic eruptions and are composed of volcanic ash and pumice particles that are bonded by natural cements or are naturally welded; they make excellent building materials and have been proposed as a medium for industrial and nuclear waste storage. Tuff deposits are often hundreds of meters thick and can cover hundreds to thousands of square kilometers. The most common tuffs used for building material are ignimbrites (pyroclastic flow deposits), in which the pumice and glass shards have been sintered by heat immediately after deposition or have been bonded by natural cements precipitated from fluids percolating through the deposits. When used for building stone, ignimbrite is sawn or broken away from a quarry face along natural cooling joints and then fashioned into blocks by hand or with power saws. These blocks, with enough strength for multiple-story buildings, stone walls, and other structures, are resistant to weathering, are lightweight, and have good insulating properties—better than most other natural building stones.

Geological Society of America
Special Paper 408
2006

Chapter 1

Introduction

Grant Heiken

331 Windantide Place, Freeland, Washington 98249-9683, USA

Many cities in volcanic regions have serious problems with industrial waste that necessitate an understanding of tuff hydrology (Table 1.1). Determining the extent and location of wastes in tuff and locating the pathways along which they flow require an understanding of the medium—tuff. Unfortunately, there is little published information on the hydrology of tuff deposits, applied either to evaluation of water resources or to waste remediation.

This book is the product of "Workshop on Tuffs—Their Properties, Uses, Hydrology, and Resources," held in Santa Fe, New Mexico, USA, 10–15 November 1996 (sponsored by the Los Alamos National Laboratory, The Yucca Mountain Project, and UNESCO).

The purpose of the workshop was to bring together a multidisciplinary group of scientists and engineers to integrate academic and applied research on tuffs and to initiate new collaborations. At this time, most of the appropriate data and publications on the physical, geological, hydrologic, and engineering properties are scattered through journal articles and government reports; there has been very little integrated work on tuffs, especially prepared for the nongeologist. Attending the workshop were volcanologists, hydrologists, civil engineers, architects, environmental chemists, and archaeologists. They were from industry, academia, international organizations, and governments, and they came from 11 countries.

The tuff workshop covered the following topics: (1) the distribution and physical and chemical variations of tuff deposits in relation to their genesis and deposition (geology and volcanology); (2) the physical and chemical properties of tuff as a building stone and the process of quarrying tuffs (civil engineering and architecture); (3) cost-effective and earthquake-resistant construction in tuff and using tuff for indigenous housing in developing countries; this topic matches one of the goals of the International Decade for Natural Disaster Reduction (architecture and civil engineering); (4) tuff as a medium for industrial and nuclear waste storage, including

TABLE 1.1. MAJOR CITIES ON TUFF DEPOSITS
Addis Ababa, Ethiopia
Ankara, Turkey
Arequipa, Peru
Bandung, Indonesia
Ciudad de Chihuahua, Mexico
Durango, Mexico
Erzurum, Turkey
Guadalajara, Mexico
Guatemala City, Guatemala
Jerevan, Armenia
Kagoshima, Japan
Kumamoto, Japan
Kumayri, Armenia
Managua, Nicaragua
Manila, Philippines
Medan, Indonesia
Medellin, Colombia
Mexico, D.F., Mexico
Nagasaki, Japan
Napoli, Italy
Quito, Ecuador
Roma, Italy
Salta, Argentina
San Jose, Costa Rica
San Luis Potosi, Mexico
San Salvador, El Salvador
Santiago de Chile
Sapporo, Japan
Taipei, Taiwan
Tashkent, Uzbekistan
Tbilisi, Georgia
Tegucigalpa, Honduras
Yogyakarta, Indonesia

drilling in tuffs (mineralogy, chemistry, and nuclear engineering); (5) the hydrology and groundwater quality of tuffaceous aquifers (hydrology and water resources engineering); (6) uses of tuff in the past (anthropology and archaeology); (7) health problems associated with tuffs (biological, geochemical, and mineralogical aspects); (8) case studies of infrastructure problems in cities located on tuff deposits (engineering, geology,

Heiken, G, 2006, Introduction, *in* Heiken, G., ed., Tuffs—Their properties, uses, hydrology, and resources: Geological Society of America Special Paper 408, p. 3–4, doi: 10.1130/2006.2408(01). For permission to copy, contact editing@geosociety.org. ©2006 Geological Society of America. All rights reserved.

hydrology, architecture, quarrying, and environmental problems); and (9) the hazards of pyroclastic eruptions.

ACKNOWLEDGMENTS

The workshop was supported by grants from the Yucca Mountain Program and in-kind support from the Los Alamos National Laboratory. Special thanks are due to Julie Canepa, who helped make the workshop possible, and LeeRoy Herrera, who was the meeting facilitator. Reviews of the manuscript were provided by Robert Levich (now retired from the Yucca Mountain Project) and David Broxton (Los Alamos National Laboratory). Thank you!

MANUSCRIPT ACCEPTED BY THE SOCIETY 29 DECEMBER 2005

Geological Society of America
Special Paper 408
2006

Chapter 2.1

Where do tuffs fit into the framework of volcanoes?

R.V. Fisher*
University of California, Santa Barbara, California, USA

Grant Heiken
331 Windantide Place, Freeland, Washington 98249-9683, USA

M. Mazzoni*
University of La Plata, Argentina

Tuffs are consolidated pyroclastic or volcaniclastic rocks. To understand the properties of tuffs, the heart of this handbook, the processes that create and deposit them need to be understood. This brief chapter is to acquaint the reader with the basic concepts and terms associated with explosive volcanism.

A Summary of the Origins of Pyroclastic Rocks

Explosive volcanic eruptions that deposit tuff are frequent and exhibit a wide range of activity and magnitude. To understand the explosive eruption process is difficult because we can approach only the smallest and least energetic of eruptions, the lava fountains of basaltic magmas. In contrast, the world's largest explosive volcanoes can erupt hundreds or thousands of cubic kilometers of ash and pumice in a few days, devastate thousands of square kilometers around a volcano, and produce ash clouds that circle the planet—these eruption processes are difficult to investigate. Even for the smallest of eruptions, we cannot see what is going on in the vent, so how do we know how magma is fragmented and erupted explosively to produce the deposits that eventually become tuffs?

Geophysical Observations

On a few well-instrumented volcanoes, observations of long-period seismic tremor below and within volcanoes led seismologists to conclude that explosive eruptions are caused by pressure disturbances within the rising magma (e.g., Morrissey and Chouet, 1997). Morrissey and Chouet interpreted the long-period tremor, with periods of 0.2–2 s as resonances within magma-filled volcanic conduits or flow-induced nonlinear oscillations. Some of these events occur before explosive activity begins. This approach to studying explosive volcanic eruptions is still in its infancy, but may provide the observations needed to understand explosive eruption phenomena.

Theory and Numerical Modeling

An understanding of magma fragmentation began to develop in the 1930s, and major theories evolved in the 1950s and 1960s. These theories were based on rheological properties of a vesiculating magma. Magma "vesiculates" when dissolved gas comes out of solution to form bubbles because of decreasing pressure as magma nears Earth's surface. During the 1980s, papers on the theoretical behavior of vesiculating liquids and fragmenting magmas led to a new subdiscipline within volcanology—physics-based eruption-process modeling (e.g., Valentine and Wohletz, 1989). High-performance computing and computing codes originally developed for weapons research were applied to the understanding of explosive volcanism. These models demonstrated that, depending on magma composition, gas content, and magma flux, fragmentation can occur along a moving rarefaction wave, moving down the volcano conduit from the surface to depths of 2–3 km. The gas and fragmented products (ash and pumice) are subsequently accelerated out of the vent to form ash fallout and pyroclastic flows.

*Deceased.

Fisher, R.V., Heiken, G., and Mazzoni, M., 2006, Where do tuffs fit into the framework of volcanoes? *in* Heiken, G., ed., Tuffs—Their properties, uses, hydrology, and resources: Geological Society of America Special Paper 408, p. 5–9, doi: 10.1130/2006.2408(2.1). For permission to copy, contact editing@geosociety.org.

Experiments

Wohletz and McQueen (1984) conducted a series of experiments in which a melt (thermite) was mixed with differing quantities of water. When mixed in the proper proportions (H_2O:melt of 0.2 to ~10), the two fluids mixed explosively and produced finely fragmented hydrovolcanic (or hydromagmatic) ash. These ashes are typical of what we find for volcanoes associated with shallow-water bodies or very productive aquifers. Numerical modeling of magma fragmentation caused by gas coming out of solution is a current topic of active research by groups in Germany and Italy.

Studies of Gases Released from Volcanoes

Recent studies at Galeras volcano, Colombia (e.g., Stix et al., 1997) and Popocatépetl volcano in Mexico have shown that decreases in the flux of gases released from the volcano may precede explosive eruptions of domes slowly growing within the crater. An understanding of the volume and composition of gases within a magma and how they are released to the atmosphere is crucial to understanding explosive volcanic processes.

Laboratory Studies of the Products of Explosive Volcanic Eruptions

Volcanic ash and pumice are essentially bits of fragmented foam formed as gases come out of solution when confining pressure drops as magma approaches Earth's surface. Early theories concluded that explosive activity occurred when all the vesicles (gas bubbles) coalesced. This is partly true, but if all the gas bubbles were to coalesce, there would be no pumice or scoria. By studying bubble walls (shards), pumice and scoria, it is possible to relate their shapes to the viscosity of the magma, the amount of gas within that magma, and estimate the degree of magma-water interaction (Heiken and Wohletz, 1985; Wohletz and Heiken, 1992).

Field Observations

Fragmentation processes that form pyroclasts can be evaluated by studying older volcanoes that have been eviscerated by erosion or by drilling into the conduits of relatively young volcanoes. Heiken et al. (1988) studied cores from drill holes adjacent to young rhyolite domes along the Inyo Dome chain in California. They found that fractures at depths of 300 m and 380 m, which formed during the eruption, were filled by poorly vesiculated shards and fragmented host rock. These shards were similar in composition to much more vesicular pumices erupted at the surface, which suggested that much of the actual vesiculation (bubble growth) had occurred within the uppermost 400 m of the conduit.

GLOSSARY OF TERMS

The term *pyroclastic* refers to volcanic materials ejected from a volcanic vent. But there are several other ways to make volcanic particles. *Volcaniclastic* refers to explosively fragmented volcanic particles and their deposits, regardless of origin (Fisher, 1961). Volcanic particles generated by weathering and erosion are termed *epiclastic*.

Generic Types of Volcaniclastic Particles

Pyroclasts

Pyroclasts are pyroclastic particles formed by vesiculation of magma and subsequent fragmentation by brittle fracturing caused by shock waves moving down the conduit from the surface. They are ejected from a volcanic vent either in air or beneath water.

Hydroclast

Hydroclasts form by steam explosions during magma-water interactions. In many explosive eruptions, fragments are formed both by magma expansion and fracturing as described above and by explosive magma-water interactions; there is no special term for these "hybrid" clasts.

Autoclast

Autoclasts form by mechanical friction during movement of lava and breakage of cool brittle outer margins, or gravity crumbling of spines and domes.

Alloclast

Alloclasts form by disruption of pre-existing volcanic rocks by igneous processes beneath Earth's surface.

Epiclast

Epiclasts are lithic clasts and minerals released by ordinary weathering processes from pre-existing consolidated rocks. Volcanic epiclasts are clasts of volcanic composition derived from erosion of volcanic rocks.

The terms *pyroclastic, hydroclastic*, and *epiclastic* refer to the initial process of fragmentation. A pyroclast, for example, cannot transform into an epiclast by reworking with water, wind, or glacial action.

Varieties of Pyroclastic Ejecta According to Origin

Essential (or juvenile). Essential pyroclasts are derived directly from erupting magma and consist of dense or inflated particles of chilled melt, or crystals (phenocrysts), in the magma prior to eruption.

Cognate (or accessory). Cognate particles are fragmented comagmatic volcanic rocks from previous eruptions of the same volcano.

Accidental. Accidental fragments (or lithoclasts) are derived from the subvolcanic basement rocks and therefore may be of any composition.

Names of Pyroclasts and Deposits According to Grain Size

Originally, terms used to describe volcaniclastic materials were developed based on the grain size of the fragmental material. The classification followed, more or less, the size classification used by sedimentologists to describe clastic rocks.

Ash: particles <2 mm in diameter. Volcanic ash is composed of vitric, crystal, or lithic particles (rock fragments of juvenile, cognate, or accidental origin). Tuff is the consolidated equivalent of ash. Classification may be made according to environment of deposition (lacustrine tuff, submarine tuff, subaerial tuff) or manner of transport (fallout tuff, ash-flow tuff). Reworked ash (or tuff) is named according to the transport agent (fluvial tuff, aeolian tuff).

Lapilli: fragments 2 mm to 64 mm in diameter. Lapilli-size particles may be juvenile, cognate, or accidental. Lithified accumulations with >75% lapilli is *lapillistone. Lapilli-tuff* is a lithified mixture of ash and lapilli, with ash-sized particles making up 25%–75% of the pyroclastic mixture. Lapilli are angular to subrounded. Subrounded forms are commonly of juvenile origin. *Accretionary lapilli* are lapilli-size particles that form as moist aggregates of ash in eruption clouds, by rain that falls through dry eruption clouds, or by electrostatic attraction (Schumacher and Schmincke, 1991). *Armored lapilli* form when wet ash becomes plastered around a solid nucleus, such as crystal, pumice, or lithic fragments, during a hydrovolcanic eruption.

Bombs or blocks: fragments >64 mm. *Bombs* are thrown from vents in a partly molten condition and solidify during flight or shortly after they land. Bombs are exclusively juvenile. Molten clots are shaped by drag forces during flight and are modified by impact if still plastic when they hit the ground. Bombs are named according to shape—ribbon bombs, spindle bombs (with twisted ends), cow-dung bombs, spheroidal bombs, and so on. *Bread-crust bombs* are so-called because of bread-crust patterns on bomb surfaces resulting from stretching of their outer solidified shell by internal gas expansion. They are commonly from magma of intermediate and silicic compositions. *Basaltic bombs* usually show little surface cracking, although some may have fine cracks caused by stretching of a thin, glassy surface over a still-plastic interior upon impact. *Cauliflower bombs* have coarsely cracked surfaces and dense interiors caused by rapid quenching in aqueous environments. They develop during hydrovolcanic eruptions.

Blocks are angular to subangular fragments of juvenile, cognate, and accidental origin derived from explosive extrusion or crumbling of domes.

Pyroclastic breccia is a consolidated aggregate of blocks containing <25% lapilli and ash.

Volcanic breccia applies to all volcaniclastic rocks composed predominantly of angular volcanic particles >2 mm in size.

Agglomerate is a nonwelded aggregate consisting predominantly of bombs. It contains <25% by volume lapilli and ash.

Pumice, scoria, and cinders are named without reference to size, but are usually lapilli or larger sizes. Their degree of vesicularity differs.

Pumice is a highly vesicular glass foam with a density of <1 g/cm^3; bubble walls are composed of translucent glass.

Scoria (also called cinders) are mafic particles that are less inflated than pumice. They are generally composed of tachylite (basaltic glass with quench crystals).

Spatter applies to bombs, usually basaltic, that form from lava blebs that readily weld (agglutinates) upon impact and contrasts with scoria, which do not stick together. Scoria (or cinder) cones, for example, are composed largely of loose particles; spatter cones are composed mainly of agglutinated blebs or larger isolated lava tongues.

Volcanoes and Facies

A volcano is a mound, hill, or mountain constructed during the eruption of lava or volcaniclastic material—usually both. Volcanoes are grouped into a few families according to shape, size, and types of extruded volcanic material. The composition of the original magma or the effects of mixing with surface water are the principal reasons for similarities and differences. *Composite volcanoes, domes, shield volcanoes, cinder cones, calderas,* and their associated facies are described here.

Lava flows and ***volcaniclastic deposits*** are the two main facies of volcanoes. Volcaniclastic deposits are the most varied and are composed of pyroclasts, reworked pyroclasts, autoclastic pyroclasts, and epiclastic particles. Pyroclasts may be deposited as fallout deposits, pyroclastic flow deposits, and pyroclastic surge deposits. Reworked pyroclastic deposits are deposited by streams, wind, and glaciers. Epiclasts are generated by weathering that releases particles from any hard rock.

Calderas are closed volcanic depressions that form during the subsidence of large blocks in a roof zone above a partially evacuated magma chamber. These collapse craters are mostly filled with slump blocks from crater walls, ash and pumice from the eruption, and lacustrine sediments. Most calderas are larger than composite volcanoes, but are less evident because of their broad low-standing profiles.

Composite Volcanoes

Description. Andesitic to dacitic magmas construct high-standing composite volcanoes with large volumes and great heights, and therefore are important sources of sediment (Hackett and Houghton, 1989). Composite volcanoes are often associated with subduction zones.

Composite volcanoes are constructed of interbedded lava flows, pyroclastic and autoclastic materials, and reworked volcanic debris. They may be the products of several volcanic episodes. They erode rapidly and are sources of epiclastic volcanic debris. Between eruptive episodes, the volcano can be severely eroded, then rebuilt, sometimes with compositionally different products. Their slopes are made of innumerable layers of rubble derived from the breakup of brittle lava and dome rocks, pyroclastic layers, a few lava flows, and reworked volcanic debris. Composite volcanoes may have satellite domes on their flanks or in craters.

Andesitic eruptions provide abundant fragmental debris to construct the composite cones, but not all of the magma rises to the summit of a volcano. Some penetrates the fragmental layers of a volcano as dikes or as sills, which parallel the boundaries of rock layers. Multiple intrusive events therefore build a rigid framework that supports the accumulation of volcanic rubble and lava flows to heights greater than cones that do not have such a framework.

Some of the highest composite volcanoes in the world include Klyuchevskoy volcano, Kamchatka, Russia (4725 m above sea level [asl]) and Ojos del Salado (6870 m asl) and Llullaillaco (6723 m asl) on the border of Argentina and Chile. Depending upon latitude, large volcanoes create their own microclimate where seasonal winter snow and rain lead to widespread erosion of loose debris from their slopes. Debris flows that originate on the steep slopes may travel more than 100 km down valleys.

Composite volcanoes can grow to 3000 or 4000 m in height, although, commonly, the instability of the accumulated deposits results in sector collapse and debris-flow avalanches similar to that which occurred during the 1980 eruption of Mount St. Helens, USA. Debris avalanches occur when a large sector of an edifice collapses, sending material down the mountain at a high rate of speed. The cause of the collapse may be that the core of the volcano is highly altered by hydrothermal activity (Crandell, 1971; Vallance and Scott, 1997), or oversteepened by intrusion of magma into the edifice (Siebert et al., 1987). The rocks of the edifice shatter during transport, but commonly remain in approximately the same stratigraphic relation with each other as blocks within a moving shattered-rock matrix (Voight et al., 1981). As the slope decreases, the avalanche slows and deposits the blocks as hummocks of shattered rock, while the matrix between the hummocks continues on, in many cases mixing with water to form mudflows (lahars) or high-concentration flows downstream.

Facies. Composite volcanoes have complex facies relationships that can be grouped into *near-source, intermediate-source,* and *distal* facies (Fisher and Schmincke, 1984).

Near-source facies. The near-source facies of composite volcanoes consists of lava flows and volcaniclastic debris. The near-source facies includes all rock types that comprise the edifice of the volcano, including abundant poorly bedded or massive autoclastic and pyroclastic breccias. These are interbedded with coarse-grained fallout tuffs and reworked pyroclastic material. In areas where erosion has removed all but the volcanic roots, the near-source facies consists mainly of a complex of stocks, sills, and dikes and some intrusive or extrusive breccia and tuff. Widespread fumarolic alteration is common.

Fragmentation of material in the near-source facies occurs during collapse of brittle lavas, domes, and spines, and from explosive eruptions. Volcaniclastic accumulations are typically thick, wedge-shaped, and discontinuous. Deposition occurs by accumulation in alluvial fans, by slump and flow within valleys, or by fallout mantling ridges and valleys. Unconformities and deposition on irregular surfaces make establishing lateral continuity of the deposits difficult. Interbedded breccias and lava flows originate from different places at different times from the volcano. Breccias may interfinger with tuff from other volcanoes within the region. Stratigraphic continuity of contemporaneous but isolated sections can sometimes be established by identifica-

tion of distinctive fallout volcanic ash layers. More commonly, it can only be done by the piecing together several interfingering units that link the local sections to known stratigraphy, or, in many cases, it cannot be done at all (Smedes and Prostka, 1972).

Intermediate-source facies. The intermediate-source facies is commonly found in the lowlands immediately surrounding the volcanic center. Rocks include those deposited from pyroclastic flows, lava flows, fallout processes, and their reworked and erosional products, such as materials deposited by debris flows, floods, and common stream deposits. The outer edges of this facies have increasing amounts of reworked pyroclastic debris and volcaniclastic sediment.

Distal facies. The distal facies consist of fallout tephra (volcanic ash) and fine-grained epiclastic volcanic sediments in areas far from the source, much farther than lava or pyroclastic flows can travel. Its transition with the intermediate-source facies is generally lost by erosion. The distant-source facies rocks are thin, well sorted, and are commonly interbedded with nonvolcanic sediments. If deposition of ash occurs slowly over thousands of years, ash sequences may become very thick, as is the case for the Miocene John Day Formation of central Oregon (Fisher and Rensberger, 1973).

Domes

Description. Volcanic domes may occur as individual volcanoes or as dome-like protrusions of lava in craters or on the flanks of composite cones. Lava domes result from the slow extrusion of highly viscous, usually silica-rich magma. Domes are generally small, but some are volcanoes that stand alone and exceed 25 km³ in volume. Many dome-producing eruptions begin explosively and form explosion craters rimmed with pyroclastic debris. Explosive activity wanes as gas content and pressure within the magma decreases. With lowered gas pressures, the magma extrudes slowly as viscous rhyolite lava that forms thick stubby flows or bulbous domes (Fink, 1987). As a lava dome grows, steep margins become covered with sharp glassy blocks of rubble. Domes on steep slopes may collapse as a mass of hot rubble that fragments into pyroclastic flows.

Facies. Facies around a dome mimic those around composite cones, but at a highly reduced scale. A near-vent facies consists of massive lava and the volcaniclastic rubble that surrounds the lava mound. Most of the volcaniclastic rubble that buries domes is formed of autoclastic debris. Such debris also forms from the collapse of the fronts of advancing silicic lava flows and the collapse of spines that tumble down the surface of steep-sided composite volcanoes. Intermediate facies are similar to those of a stratovolcano in those cases in which domes are explosive or have lava flows distinct from the source volcano. The autoclastic debris is interbedded with volcaniclastic deposits of nearly all origins that can occur on the sides of composite volcanoes. Where domes are isolated, the debris is monolithologic and massive.

Shield Volcanoes

Description. Shield volcanoes are broad, have low slopes, and are composed almost exclusively of solidified layers of basaltic lava flows. The fluid lavas can flow long distances, and therefore construct gentle slopes and broad summit areas, unlike the steeper-sided composite volcanoes or cinder cones. Some hydroclastic materials are formed by interaction with magma and water in the vent, as occurred at Kilauea in 1924. Scoria (cinder) cones, composed of basaltic tephra, occur on the surface of some shield volcanoes.

Facies. Shield volcanoes have a simple and distinct facies pattern, namely, layer after layer of thin basalt flows with the geometry of interlaced tongue-shaped bodies, some of which may have flowed 30 km. Unconformities marked by soil zones are common. Pyroclastic deposits consist of Pele's hair and tears, and hydrovolcanic tephra are rare.

Scoria (Cinder) Cones

Description. Scoria (cinder) cones are relatively small volcanoes composed of basaltic fragments. They are one of the most common volcanic landforms on Earth. Scoria clasts are small, nut- to fist-sized or larger pieces of lava containing abundant vesicle cavities that cooled during or soon after Strombolian eruptions (explosive bursts of molten and semi-molten clasts). Scoria cones are usually constructed completely of fragmental cinders and form steep-sided mounds with a small crater at the top. Scoria pyroclasts are deposited ballistically until the angle of repose is reached, when the loose material avalanches to the base of the cone. Most cinder cones are composed of interbeds of ballistic fallout and avalanche deposits. Very close to the vent, the cinders are often welded because they were deposited while still hot and soft. Lava flows commonly break through the lower slopes or overflow from lava lakes in the summit crater. Scoria cones may occur alone but commonly are in clusters, sometimes associated with more than 100 neighboring cinder cones. They also occur on the slopes of shield volcanoes.

Facies. Scoria cones are local features composed of lenticular and wedge-shaped red or black beds of bombs, lapilli-sized scoria, and ash. Hawaiian cinder cones commonly consist of bombs, lapilli consisting of Pele's tears, and lesser amounts of ash. In fields where there are many cinder cones, lava and basaltic tephra are interbedded.

Calderas

Description. A caldera is a very large collapse crater formed during collapse of the ground surface during an eruption. Following the extrusion of such large volumes of ash, pumice, and rock emptied from the magma chamber, the ground above the magma chamber collapses into the resulting void. The dimensions of calderas are quite variable, from a few kilometers to as large as 60 km in diameter. The large ones are easily visible from space, but on the ground they may be difficult to recognize because their configuration is not visible from a single viewing place. Caldera-forming eruptions usually produce widespread pyroclastic flow deposits and abundant ash fallout deposits. Calderas can also form during withdrawal of basaltic magma that is erupted along the flank of a shield volcano. The collapse occurs without any eruption at the volcano summit; an excellent example of this is Halemaumau caldera, Kilauea volcano, Hawaii.

Facies. Calderas, as well as composite volcanoes, can produce enormous amounts of volcaniclastic debris. Unlike composite volcanoes, calderas have large-diameter craters generally without high-standing edifices, with correspondingly lower rates of erosional reworking of deposits. Caldera collapse (Smith and Bailey, 1968) produces thick deposits that accumulate within the crater as an *intra-caldera facies*. The intra-caldera facies within the subsided area includes ignimbrite deposits hundreds of meters thick. With resurgence, the resulting moat is filled by pyroclastic rocks, lava flows, lake sediments, epiclastic volcanic sediments, and by landslide or talus breccias from the caldera wall. The caldera-outflow facies consist of ignimbrite sheets extending many kilometers from the caldera.

REFERENCES CITED

Crandell, D.R., 1971, Postglacial lahars from Mount Rainier volcano, Washington: U.S. Geological Survey Professional Paper 677, 75 p.

Fink, J.H., ed., 1987, The emplacement of silicic domes and lava flows: Geological Society of America Special Paper 212, 145 p.

Fisher, R.V., 1961, Proposed classification of volcaniclastic sediments and rocks: Geological Society of America Bulletin, v. 72, p. 1409–1414. Reprinted in Benchmark Papers in Geology, Sedimentary Rocks: Concepts and History, Carozzi, A.V., ed., Halsted Press, p. 220–225, 1975.

Fisher, R.V., and Rensberger, J.M., 1973, Physical stratigraphy of the John Day Formation, central Oregon: University of California Publications in Geological Sciences, v. 101, 33 p.

Fisher, R.V., and Schmincke, H.-U., 1984, Pyroclastic rocks: Berlin, Springer-Verlag, 472 p.

Hackett, W.R., and Houghton, B.F., 1989, A facies model for a Quaternary andesitic composite volcano: Ruapehu, New Zealand: Bulletin of Volcanology, v. 51, p. 51–68, doi: 10.1007/BF01086761.

Heiken, G., and Wohletz, K., 1985, Volcanic ash: Berkeley, University of California Press, 246 pp.

Heiken, G., Wohletz, K., and Eichelberger, J., 1988, Fracture fillings and intrusive pyroclasts, Inyo Domes, California: Journal of Geophysical Research, v. 93, p. 4335–4350.

Morrissey, M.M., and Chouet, B.A., 1997, A numerical investigation of choked flow dynamics and its application to the triggering mechanism of long-period events at Redoubt volcano, Alaska: Journal of Geophysical Research, v. 102, p. 7965–7983, doi: 10.1029/97JB00023.

Schumacher, R., and Schmincke, H.-U., 1991, Internal structure and occurrence of accretionary lapilli; a case study at Laacher See volcano: Bulletin of Volcanology C, v. 53, p. 612–634.

Siebert, L., Glicken, H., and Ui, T., 1987, Volcanic hazards from Bezymianny- and Bandai-type eruptions: Bulletin of Volcanology, v. 49, p. 435–459.

Smedes, H.W., and Prostka, H.J., 1972, Stratigraphic framework of the Absaroka Volcanic Supergroup in the Yellowstone National Park region: U.S. Geological Survey Professional Paper, v. 729-C, p. C1–C33.

Smith, R.L., and Bailey, R.A., 1968, Resurgent cauldrons: Geological Society of America Bulletin, v. 116, p. 613–662.

Stix, J., Calvache, V.-M., and Williams, S.N., eds., 1997, Galeras volcano, Colombia; interdisciplinary study of a decade volcano: Journal of Volcanology and Geothermal Research, v. 77, p. 1–4.

Valentine, G.A., and Wohletz, K.H., 1989, Numerical models of Plinian eruption columns and pyroclastic flows: Journal of Geophysical Research, v. 94, p. 1867–1877.

Vallance, J.W., and Scott, K.M., 1997, The Osceola mudflow from Mount Rainier: Sedimentology and hazards implications of a huge clay-rich debris flow: Geological Society of America Bulletin, v. 109, p. 143–163, doi: 10.1130/0016-7606(1997)109<0143:TOMFMR>2.3.CO;2.

Voight, B., Glicken, H., Janda, R.J., and Douglass, P.M., 1981, Catastrophic rockslide avalanche of May 18: U.S. Geological Survey Professional Paper, v. 1250, p. 347–348.

Wohletz, K., and Heiken, G., 1992, Volcanology and geothermal energy: Berkeley, University of California Press, 432 p.

Wohletz, K.H., and McQueen, R.G., 1984, Experimental studies of hydromagmatic volcanism, *in* Explosive volcanism: Inception, evolution, and hazards: Washington, National Academy Press, Studies in Geophysics, p. 158–169.

MANUSCRIPT ACCEPTED BY THE SOCIETY 29 DECEMBER 2005

Geological Society of America
Special Paper 408
2006

Chapter 2.2

Tuff mineralogy

David Vaniman
Los Alamos National Laboratory, Los Alamos, New Mexico 87545, USA

POSTDEPOSITIONAL PROPERTIES

Crystallization and Alteration

The defining constituent of any magma is molten rock, but few magmas are entirely liquid. Magmas are generally erupted with some proportion of crystals already forming from the melt or entrained from complex mixing or assimilation events that precede eruption. These crystals, supplemented or depleted by interaction with the melt after eruption, generally form the group of minerals that are visible as phenocrysts in the erupted rock, which are the largest of the crystals formed and those visible with the unaided eye or with only slight magnification.

Phenocrysts in tuffs are often accompanied by rock (lithic) fragments that may be derived by the mechanical plucking of rock from the magma chamber walls or the conduit, or by incorporation of debris from the surface by pyroclastic density currents. The minerals in this suite of phenocrysts and lithic fragments can form anywhere from <1% to several tens of percent of the tuff. In tuffs that are termed crystal-rich or lithic-rich because of the abundance of these constituents, the variety of minerals of such early origin is an important factor in determining later alteration and subsequent modes of water-rock interaction. However, the smaller crystals that form by alteration of the larger volume of erupted glass are almost always more abundant and have much greater surface area, making the mineralogy of alteration the most important factor in determining those tuff properties that depend on access to mineral surfaces. This includes virtually all forms of water-rock interaction. The processes that lead to alteration of tuff glass are described herein under the sequential posteruption categories of devitrification, vapor-phase alteration, and alteration to clays and zeolites. All tuffs, if exposed to shallow hydrothermal alteration or to burial and metamorphism, can undergo further alteration that supplements or overprints the mineralogy of these earlier alterations. This topic is dealt with last.

Devitrification

Volcanic glass is abundant in freshly erupted tuffs, both as ash and pumice. Devitrification is a term used almost exclusively to refer to the crystallization of these glasses soon after eruption, during the slow cooling of thick tuff deposits. Although other processes lead to loss of glass (vapor-phase or clay and zeolite alteration), the term devitrification principally describes this early high-temperature crystallization of glass. This first stage of glass loss results in the formation of abundant finely crystalline feldspars and silica minerals, with only very minor amounts of other mineral types. Feldspar (Ca, Na, K, Al-silicate) predominates because of the close resemblance of siliceous glass to the alkali-aluminosilicate composition of Na, K-feldspar. The silica minerals are less abundant but comprise most of the remaining devitrification products. In tuffs, all three of the principal silica minerals may occur (quartz, cristobalite, and tridymite, each with a different crystal structure but with the same simple formula, SiO_2).

For all igneous rocks, mineral growth can be calculated in terms of a normative mineral composition, indicating the relative abundances (wt%) of minerals that form from the melt on crystallization. For tuff glasses, this type of calculation can also serve as an "ideal" indication of the ratios of minerals that may form from the glass upon devitrification. As Table 2.2.1 shows, a wide variety of siliceous glasses can crystallize to form intergrowths with between 97% and 82% feldspar plus silica minerals. The most voluminous of large-scale tuff eruptions, accounting for many of the terranes in which tuffs are significant in landform development, are in the 65%–77% SiO_2 range, with greater than 90% normative feldspar plus silica minerals.

Other tuff compositions can occur; rocks that are described as "undersaturated" with respect to silica can form little or none of the silica minerals and instead form feldspar-like minerals that are silica poor (such as nepheline). Such tuffs are important in some parts of the world, but the vast majority of tuffs have

Vaniman, D., 2006, Tuff mineralogy, *in* Heiken, G., ed., Tuffs—Their properties, uses, hydrology, and resources: Geological Society of America Special Paper 408, p. 11–15, doi: 10.1130/2006.2408(2.2). For permission to copy, contact editing@geosociety.org.

glasses that, when devitrified, form the feldspar plus silica mineral assemblage described here.

It is possible to obtain the wrong impression about importance of mineral types from Table 2.2.1. Minor minerals that contain Fe, Ti, Mn, Mg, and P are much more diverse than this table suggests. For example, the constituents simply represented as "pyroxene" in reality may include amphiboles and biotite; the oxides typically include magnetite and ilmenite. Moreover, these minor minerals contain the heavy metals that are important in determining critical aspects of some water compositions, as groundwater systems develop in aging tuffs. However, the most abundant of these lesser minerals and the ones most accessible to invading water are those that form after devitrification, through the processes of vapor-phase alteration or alteration to clays and zeolites.

Vapor-Phase Alteration

In the interiors and particularly in the upper portions of thick tuff deposits, water vapor released from devitrifying glass, often supplemented with meteoric water, provides a chemically active and very mobile system for redistributing constituents within the tuff. Silicic magmas rising rapidly from depths of ~10 km can contain up to 8% H_2O. Most of this water is lost as pressure drops during eruption, to the point where the pyroclastic glasses that are products of the eruption contain <1% H_2O. This water is released as a gas as the anhydrous feldspars and silica minerals form during devitrification; the water available from devitrifying glass, plus a component of entrapped water from the volatile constituents that drove the eruption provide the major part of the fluid involved in vapor-phase alteration. Additional gases (principally CO_2, S, and halogen species) contribute to the vapor phase. In order for vapor-phase alteration to occur, the water-to-rock ratio must not exceed certain limits, or the glass is quenched and altered by other processes. In most phreatomagmatic eruptions and in marine or lacustrine tuff deposits, the hot vapor is quenched by the abundant surface water or groundwater and alteration proceeds directly to other modes, principally clay formation.

The volume of the tuff is also important in determining whether or not vapor-phase alteration takes place. In Plinian and in small-scale pyroclastic-flow eruptions, most vapor is lost during eruption, deposition, and subsequent flow compaction. In larger ash flows, the vapor is trapped, and the vapor-to-rock volume ratio often increases as the glass devitrifies. This vapor buildup can result in the development of gas expansion features (lithophysal cavities and fracture systems) that form as the surrounding tuff is compacted and welded under its own heat and mass. Vapor-phase alteration is thus rare in tuffs that are less than a few tens of meters thick.

Vapor-phase mineralogy includes the feldspars and silica minerals that are the basic components of devitrification, but the full complement of vapor-phase minerals is much more complex. Complexity arises mainly from the many components carried by the vapor. The devitrification feldspars and silica minerals incorporate only Ca, Na, K, Al, Si, and O in abundance, whereas the vapor-phase minerals include hydrous silicates, oxides, and other minerals, which may include Fe, Mn, Mg, S, F, Cl, and a host of trace metals that are expelled from the crystallizing glass as they are excluded from the structure of devitrification minerals. Vapor-phase mineralogy is sufficiently complex that full characterization is difficult for any tuff so altered. Example listings of vapor-phase minerals for two rhyolitic tuffs are given in Table 2.2.2.

The differences in vapor-phase mineralogy between the examples in Table 2.2.2, the Topopah Spring and Bandelier Tuffs, are striking. Causes of such differences remain poorly understood. Perhaps one of the major factors in accounting for

TABLE 2.2.1. CHEMICAL COMPOSITIONS AND NORMATIVE MINERAL COMPOSITIONS OF FOUR IGNEOUS ROCKS

	rhyolite (wt%)	rhyodacite (wt%)	dacite (wt%)	andesite (wt%)
SiO_2	76.21	72.19	67.73	62.74
TiO_2	0.07	0.33	0.50	0.56
Al_2O_3	12.58	12.62	15.44	16.53
Fe_2O_3	0.30	3.14	0.69	1.71
FeO	0.73	1.12	2.40	2.14
MnO	0.04	0.05	0.06	0.07
MgO	0.03	0.58	1.30	3.24
CaO	0.61	2.07	3.35	6.20
Na_2O	4.05	3.45	3.85	4.08
K_2O	4.72	3.70	3.25	1.18
P_2O_5	0.01	0.02	0.15	0.16
rest	0.52	0.80	1.15	1.31
silica minerals	33.1	33.2	23.0	17.4
feldspar	64.4	59.0	66.7	64.8
pyroxene	1.2	2.3	6.8	12.5
oxides	0.6	3.5	1.8	3.5
apatite	0.0	0.2	0.3	0.4

TABLE 2.2.2. COMPARISON OF KNOWN VAPOR-PHASE MINERALS IN THE TOPOPAH SPRING TUFF, NEVADA, AND THE BANDELIER TUFF, NEW MEXICO

	Topopah Spring Tuff	Bandelier Tuff*
silicate minerals	tridymite cristobalite quartz alkali feldspar Mn-amphibole Mn-biotite Mn-garnet	tridymite cristobalite alkali feldspar zircon titanite monazite
oxide minerals	hematite pseudobrookite	magnetite (Pb,V,Ca,K) and W oxides
other minerals	fluorite	barite apatite sylvite galena Fe-sulfide $PbCO_3$ BiCl AgCl

*From Stimac et al. (1996).

these differences is the degree of welding and devitrification. In the densely welded Topopah Spring Tuff, devitrification is pervasive and was apparently complete before vapor-phase alteration began; the highly oxidizing vapor phase had developed a lithophysal fracture system for communication. This communication system (possibly fumarolic) may have allowed some vapor to escape as fluids, which carried off many of the soluble constituents that were retained as "other minerals" in the Bandelier Tuff. The more poorly welded Bandelier Tuff is much more pervasively vapor-phase altered and lacks the lithophysal fracture system of the Topopah Spring Tuff, suggesting a "stewpot" system in which most vapor constituents were retained. Such conclusions remain speculative without further data on vapor-phase systems. Nevertheless, the potential inventory of soluble minerals, particularly those containing heavy metals that may end up in subsequent groundwater systems, is an important aspect of tuff mineralogy.

Alteration to Clays and Zeolites

Thin tuff deposits and the margins of larger deposits typically retain glass zones that are unaffected by either devitrification or vapor-phase crystallization. These glasses are often subjected to much slower alteration, as the unstable glass reacts with low-temperature waters to form clays and zeolites. Clays and zeolites are also common alteration products of "wet" tuffs, which are emplaced in lacustrine or marine environments. In many tuffs, there is evidence for a general sequence beginning with the formation of clay (principally smectite) by the reaction of glass with water of moderate pH and low ionic strength and progressing to the formation of zeolites as the solution evolves to higher pH and higher ionic strength. However, there are many exceptions to this simple model, and the complexity of tuff depositional environments provides for considerable variety.

Smectites are almost universal as alteration products of silicic tuffs, but their abundance is typically a few percent; abundances >50% are not common except at hydrologic barriers in tuffs or in hydrothermally altered tuffs (see Hydrothermal Alteration and Metamorphism section). In contrast, zeolite abundances can often be >75% in altered tuffs. Zeolitic alteration generally occurs near or below the water table, but may also occur in fractures, or at hydrologic barriers in unsaturated tuffs, as is the case with smectites. Field studies recognize two major types of environments in which zeolites form in tuffs: "open-system" zeolitization of glassy layers within thick tuff accumulations (generally restricted to tuff accumulations at least several hundred meters thick), and "closed-system" zeolitization of tuffs that have been deposited in saline lake environments. Examples of both types of environments are described in the following paragraphs.

Open-system tuff zeolitization provides some of the largest zeolite accumulations in the world. The description of these as "open-system" zeolite occurrences recognizes the role of progressive groundwater modification and geochemical migration during zeolitization. Hay and Sheppard (1977) described the general reaction in the John Day Formation of central Oregon as:

TABLE 2.2.3. COMPOSITIONS OF ZEOLITES COMMONLY FORMED IN OPEN-SYSTEM TUFF ALTERATION

Zeolite	Ideal formula	Si/Al ratio
Chabazite	$(Na_2,Ca)_6(Al_{12}Si_{24})O_{72} \cdot 40H_2O$	2.0
Clinoptilolite	$(Na_4K_4)(Al_8Si_{40}O_{96}) \cdot 24H_2O$	5.0
Mordenite	$(Na,K)_8(Al_8Si_{40}O_{96}) \cdot 24H_2O$	5.0
Phillipsite	$(Na,K)_{10}(Al_{10}Si_{22}O_{62}) \cdot 20H_2O$	2.2

$$glass + Ca + Mg + H_2O \rightarrow clinoptilolite + Si + Na + K + Fe.$$

The specification of the zeolite clinoptilolite in this reaction recognizes the most common of zeolite minerals that occurs in open-system zeolitization of silicic tuffs. This is by no means the only zeolite formed by such reactions in tuffs. Zeolites have broadly overlapping compositions, and the conditions leading to the formation of a particular zeolite in a given open-system alteration scheme is still poorly understood. Table 2.2.3 summarizes the principal chemical constituents of zeolites formed in open-system alteration; the general chemical similarity between these zeolites is evident in this table.

The similarities between these zeolites are even greater when it is realized that the alkali and alkaline-earth elements in the first part of each formula are readily exchangeable. This effect arises from the open channels in the crystal structures of these zeolites. Each of the four zeolites in Table 2.2.3 can accept Na, K, and Ca, as well as many other cations in these channels. Thus, the formulae in Table 2.2.3 are only approximations, and the compositional overlap is even greater than shown. However, there are some restrictions on the occurrence of zeolites in open-system alteration that pertain to the proportions of Al and Si in the tetrahedral structural elements that form the channels in which the exchangeable cations reside. These restrictions arise from variation in Si/Al ratios and are evident in the types of zeolites that form in open-system alteration of tuffs with different silica compositions (Table 2.2.4).

From Table 2.2.4, it can be seen that high-silica zeolites are common only in tuffs with commensurately high silica content. Although the less-siliceous zeolites can also occur in these high-silica tuffs, they are less common and are often found in alteration environments where silica activity may be lower than typical. The high-silica zeolites clinoptilolite and mordenite are almost never found in basaltic tephra, where silica activity during alteration is generally too low to stabilize these zeolites.

Closed-system zeolitization provides a very different distribution of minerals. Saline lake basins have an evaporative history or undergo evaporative cycles that lead to a concentric pattern of mineral formation, with those minerals requiring the highest ionic strength and highest pH at the center of the concentric pattern. From the rim of the alteration basin to the center, the typical alteration sequence is shown in Table 2.2.5.

As in the case of open-system zeolitization, a wide variety of zeolites with exchangeable cations can form in the outer alteration

TABLE 2.2.4. MOST COMMON ZEOLITES IN TUFFS AND TEPHRAS
OF DIFFERENT SILICA CONTENT

	Clinoptilolite(5.0)	Mordenite(5.0)	Chabazite(2.0)	Phillipsite(2.2)
High-silica tuffs	X	X	X	X
Low-silica tuffs			X	X
Basaltic tephra			X	X

Note: Structural Si/Al ratio for each zeolite shown in parentheses.

TABLE 2.2.5. ZONATION OF ALTERATION IN CLOSED-SYSTEM ZEOLITIZATION
OF TUFFS DEPOSITED IN SALINE-LAKE BASINS

Outer basin margins	$\rightarrow$	$\rightarrow$	Central basin
glass	chabazite clinoptilolite erionite phillipsite mordenite	analcime	K-feldspar

Note: After Sheppard and Gude (1973).

zone, often in multiple associations. Which one of these zeolites forms depends on many factors, including Si/Al ratio, salinity, and cation composition of the brackish water. In contrast to this variety in zeolites with exchangeable cations, all of these zeolites react with waters of increased salinity and alkalinity to form the relatively simple zeolite analcime ($NaAlSi_2O_6 \cdot H_2O$). In the most saline and alkaline remnants of the central evaporative basin, where the activity of water is sufficiently diminished, any of the zeolites may react with the concentrated brine to produce low-temperature K-feldspar (adularia).

Hydrothermal Alteration and Metamorphism

Tuffs are typically deposited in tectonically unstable regions, where geothermal activity is high, and burial to depths where metamorphism is effective can occur in relatively short time spans. In these cycles of tectonic activity, the tuffs that are hydrothermally altered or metamorphosed at depth may be raised and exhumed by erosion, leaving them accessible at Earth's surface. In most cases, however, these generally deeper alteration zones are accessed by mining or drilling.

Hydrothermal alteration of tuffs can result in significant introduction or loss of chemical constituents. Bish and Aronson (1993) described albitization of deep tuffs in a fossil (11 Ma) hydrothermal system at >1.5 km depth at Yucca Mountain, Nevada. There, albite occurs in the groundmass, in veins, and accompanying calcite in pseudomorphs of feldspar phenocrysts. These minerals are associated with abundant illite with ordered structures suggesting fossil temperatures as high as 275 °C, and by lesser amounts of pyrite and rare barite. A more extensive alteration series is seen in the Jemez Mountains of New Mexico, where an active hydrothermal system occurs in tuffs within the Valles caldera, and temperatures of 210 °C are attained at a depth

of 0.5 km (Goff et al., 1989). There, the alteration extends to the surface, with a quartz-sericite zone of intense alteration overlying a chlorite-sericite zone of moderate alteration. Alteration minerals associated with the upper zone include pyrite, fluorite, and molybdenite; alteration minerals associated with the deeper zone include calcite and pyrite. This inversion of alteration intensity provides an illustration of the importance of history as well as local geology in determining the nature of hydrothermal alteration in tuffs. In the Valles caldera, it is evident that shallow alteration at temperatures >200 °C was important in some areas in the past. Both the Yucca Mountain and the Valles examples show that hydrothermal alteration of tuffs must be considered in time as well as in locale. Each occurrence of hydrothermally altered tuff must be evaluated independently.

In contrast to the highly variable mineralogy of hydrothermal alteration, the effects of burial metamorphism are more readily described in terms of a generally applicable model. The highly reactive glasses in thick sequences of buried volcanic rocks, including tuffs, have historically provided the definitive understanding of processes that bridge the gap between diagenesis and metamorphism. A simple but widely applicable sequence for low-grade burial metamorphism of rhyolite-to-dacite volcanic rocks was developed by Iijima and Utada (1971) and is summarized in Table 2.2.6.

The depth ranges for these zones and, to a lesser extent, the temperature limits, may vary significantly in other locales, but the general sequence is typical of most thick, buried tuff sequences, whether in marine or continental depositional environments.

REFERENCES CITED

Bish, D.L., and Aronson, J.L., 1993, Paleogeothermal and paleohydrologic conditions in silicic tuff from Yucca Mountain, Nevada: Clays and Clay Minerals, v. 41, p. 148–161.

TABLE 2.2.6. BURIAL METAMORPHISM ZONES OF IIJIMA AND UTADA (1971)
FOR THE NIIGATA OIL FIELD

Zone	Alteration mineralogy	Maximum temperature (°C)	Maximum depth (km)
Zone I	Fresh glass preserved	41–49	2
Zone II	Alkali clinoptilolite	55–59	2.5
Zone III	Clinoptilolite + mordenite	84–91	3.7
Zone IV	Analcime	120–124	4.6
Zone V	Albite	?	?

Goff, F., Gardner, J.N., Baldridge, W.S., Hulen, J.B., Nielson, D.L., Vaniman, D., Heiken, G., Dungan, M.A., and Broxton, D., 1989, Volcanic and hydrothermal evaluation of Valles caldera and Jemez volcanic field: New Mexico Bureau of Mines & Mineral Resources Memoir, v. 46, p. 381–434.

Hay, R.L., and Sheppard, R.H., 1977, Zeolites in open hydrologic systems, *in* Mumpton, F.A., ed., Mineralogy and geology of zeolites: Mineralogical Society of America Short Course Notes, v. 4, p. 93–102.

Iijima, A., and Utada, M., 1971, Present-day zeolitic diagenesis of Neogene geosynclinal deposits in Niigata oil field, Japan: Molecular Sieve Zeolites—I, Advances in Chemistry Series, v. 101, p. 342–349.

Sheppard, R.A., and Gude, A.J., III, 1973, Zeolites and associated authigenic silicate minerals in tuffaceous rocks of the Big Sandy Formation, Mohave County, Arizona: U.S. Geological Survey Professional Paper 830, 36 p.

Stimac, J., Hickmott, D., Abell, R., Larocque. A.C.L., Broxton, D., Gardner, J., Chipera, S., Wolff, J., and Gauerke, E., 1996, Redistribution of Pb and other volatile trace metals during eruption, devitrification, and vaporphase crystallization of the Bandelier Tuff, New Mexico: Journal of Volcanology and Geothermal Research, v. 73, p. 245–266.

MANUSCRIPT ACCEPTED BY THE SOCIETY 29 DECEMBER 2005

Geological Society of America
Special Paper 408
2006

Chapter 2.3

Fractures in welded tuff

Kenneth Wohletz
Los Alamos National Laboratory, Los Alamos, New Mexico 87545, USA

INTRODUCTION

A notable feature and, in many cases, a distinguishing aspect of welded tuff is the occurrence of numerous fractures. These fractures, generally oriented in a nearly vertical fashion, are rather equally spaced, and show little or no displacement. Produced by cooling and contraction perpendicular to isotherms in tuff after its emplacement and during its welding, these fractures are often referred to as *cooling joints*, resembling columnar joints developed in basaltic lava flows. Notable exceptions are where joints form plumose patterns in fumarolic zones. These fractures play a dominant role in the structural integrity of welded tuffs, their hydrology, and, by secondary processes, their mineralogy.

The Bandelier Tuff of New Mexico provides field data that illustrate fracture characteristics. Erupted 1.13 million years ago, the upper member of the Bandelier Tuff (Tshirege Member) is well exposed in numerous canyons that cut the Pajarito Plateau in northern New Mexico. It displays prominent fractures, which likely play an important role in the vadose-zone hydrology of the tuff and the surface manifestations of the tectonic fabric. Detailed canyon wall maps, discussed here, record the location and morphology of over 5000 fractures, represented in nearly 8 km of canyon-wall exposure. Being best developed in more strongly-welded zones, fractures extend from the surface of the tuff to the pumice fallout underlying it. The average fracture spacing is 1.5 m in studied locations giving a linear density of 65 fractures per hundred meters. Notable increases in fracture density up to 230 per hundred meters occur over tectonic lineaments associated with the Pajarito fault system. Fracture strikes are widely dispersed, but do show a crude bimodal distribution that defines a conjugate system of northwesterly- and northeasterly-oriented fracture sets. Since fracture maps represent vertical cross sections along generally west-to-east canyon walls, a fracture set generally paralleling these canyons is not well documented, but its presence is indicated by simple trigonometric approximations.

Rare surface exposures of fracture sets show a polygonal, often rhombohedral, pattern, though in places the pattern is nearly orthogonal or hexahedral. Most fractures are steeply dipping (~80°), but nearly horizontal fractures are also evident. They display both planar (constant aperture) and sinuous (variable aperture) surfaces, averaging 0.7–1.0 cm aperture in all studied areas; fractures in tectonic zones, however, show average apertures up to 5 cm. Although most fractures are not filled at depth, they are packed with detritus and secondary minerals within about 15 m of the surface. Combining linear density and fracture aperture data shows that an average of 0.7 m of total aperture exists over any 100-m interval, but it can rise to over 5 m of total aperture per hundred meters over tectonic zones.

An interesting result of this characterization is a demonstration of how a fractured-welded tuff can conceal faults by accommodating strain incrementally in each fracture over a wide area. Calculations based on the fracture data indicate that the Bandelier Tuff conceals fault displacement of up to several meters. The occurrence and distribution of fractures in welded tuff is an important consideration for slope stability and infiltration of surface water and contaminants. The development of new dual-porosity/dual-permeability numerical procedures for predicting contaminant dispersal relies on accurate input of fracture characteristics. Providing obvious pathways for dispersal, the Bandelier Tuff shows that fracture fillings of detrital materials and secondary minerals actually block migration along fractures near the tuff's surface. At deeper levels, fractures make the tuff very permeable and capable of containing large quantities of water.

BACKGROUND

Most fractures in welded tuff can be attributed to what is most widely known as columnar jointing, a textural feature displayed by generally vertical, evenly spaced cracks that intersect to form polygonal columns. These fractures are in many ways similar to

Wohletz, K., 2006, Fractures in welded tuff, *in* Heiken, G., ed., Tuffs—Their properties, uses, hydrology, and resources: Geological Society of America Special Paper 408, p. 17–31, doi: 10.1130/2006.2408(2.3). For permission to copy, contact editing@geosociety.org. ©2006 Geological Society of America. All rights reserved.

5- and 6-sided columnar joints formed in basaltic lava flows. Ross and Smith (1961) attributed columnar jointing in welded tuffs to cooling tension, but noted that the columns more frequently form rectilinear joint sets. Since that study there has been little work on characterizing these joints until hydrogeologic studies of welded tuffs started at the Nevada Test Site (NTS) in the 1970s. Barton and Larsen (1985) studied NTS joint fracture patterns in order to better understand water infiltration and movement in the vadose zone. They suggested a fractal nature to the occurrence of these fractures. More recently Fuller and Sharp (1992) characterized tuff fractures and their hydrologic properties with a detailed characterization of fracture spacing and orientation, combined with an analysis of fracture permeability.

This chapter reviews welded tuff fractures, their origins, characteristics, and possible significance. Table 2.3.1 outlines the ranges of welded tuff physical properties and shows that most notable variations (e.g., permeability) are linked to the degree of welding.

The columnar joints that produce fractures in welded tuffs are illustrated by example photographs shown in Figures 2.3.1 through 2.3.4. While nearly vertical orientations are most common, highly slanted and radiating distribution of joints are typical for tuffs that have cooled over irregular topography or have been modified by fumarolic activity. The occurrence of joints in welded tuffs is summarized by the schematic illustration shown in Figure 2.3.5.

Because most fractures in welded tuffs are manifestations of columnar jointing, characterization is achieved by measurement

TABLE 2.3.1. WELDED TUFF PHYSICAL PROPERTIES

Property	Minimum	Maximum
Bulk Density (Mg/m^3)[a]		
Nonwelded	0.5	1.8
Partially welded	1.8	2.0
Moderately welded	2.0	2.3
Densely welded	2.3	2.6
Median grain size (mm)	0.06	8
Porosity	0	0.6
Permeability (m^2)		
Nonwelded	1×10^{-17}	1×10^{-12}
Welded	1×10^{-19}	1×10^{-16}
Young's Modulus (E, Mb)	0.04	0.14
Shear Modulus (Mb)	0.02	0.6
Poisson's Ratio (ν)	0.10	0.15
Crushing strength (kb)	1.3	1.3
Cohesive strength (kb)	0.3	0.3
Compressional velocity (km/s)	0.7	4.6
Shear velocity (km/s)	0.8	2.0
Heat capacity (kJ/kg-K)[b]	1.0	1.2
Conductivity (W/m-k)	0.2	0.4
Electrical resistivity (Ω-m)		
Nonwelded	17	60
Welded	200	1400

[a] Bulk densities are dependent upon composition; density generally decreases with increasing silica content.

[b] Heat capacities for nonporous tuff can be multiplied by porosity to get effective heat capacity.

Figure 2.3.1. The Bandelier Tuff in Frijoles Canyon of northern New Mexico. Note the darker-colored, strongly welded zone near the top of the canyon wall where fractures are best displayed.

Figure 2.3.2. The Bandelier Tuff in Los Alamos Canyon showing orthogonal columns averaging ~0.5–1.0 m on a side. These columns spall from the cliff face during canyon sidewall erosion, which occurs most readily where fractures are most closely spaced.

Figure 2.3.3. Columnar jointing in the outflow facies of the Cerro Galan ignimbrite of northwest Argentina. The cliff face is ~20 m high and shows a marked curvature of the joint pattern from vertical, likely a reflection of bowed isotherms that existed in the tuff during cooling. Photograph was adapted from Francis (1993).

Figure 2.3.4. Strongly curved columnar joints in the densely welded zone of the Bishop Tuff in eastern California that radiate away from a fumarolic pipe. The joints are spaced by ~1 m. Photograph was adapted from Fisher and Schmincke (1984).

of joint spacing, orientation, and aperture. This type of characterization will be described and analyzed in this paper.

Joint Formation Model

A model for joint formation can be formulated based upon the hypothesis that jointing results from volumetric contraction and vertical compaction during tuff cooling and welding. This contraction and compaction leads to growth of in situ stresses that concentrate in excess of rock moduli. Because of their similarity to columnar joints in basaltic lavas, the model of DeGraff and Aydin (1993) should also apply to joint formation in welded tuffs. In that model, joints grow away from a cooling surface normal to maximum tensile stress.

With welding compaction, one may assume that the upper and lower surfaces of the tuff are free to contract vertically. In contrast, lateral contraction and bending of the tuff are mechanically constrained, which leads to horizontal stress buildup. The stress dissipation is strongly coupled to temperature-dependent rheology. For example, where the tuff is ductile, stresses are accommodated by viscous flow, but where it is brittle, an elastic response to stress builds up. Assuming that the brittle-ductile transition temperature (T_s) is abrupt and dependent on composition (T_s = 725–1065 °C), and rock below T_s is linearly elastic, fractures form in brittle regions and terminate where $T > T_s$, at which point the tuff responds in a ductile fashion. Because heat loss is dominantly from the bottom and top surfaces of the tuff, isotherms are horizontal where the substrate is horizontal, such that

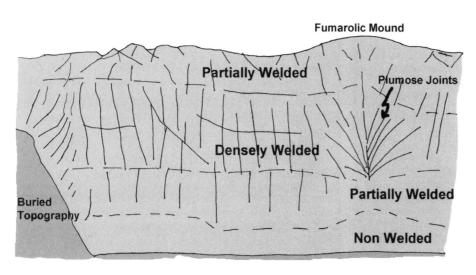

Figure 2.3.5. Schematic representation showing the distribution of columnar jointing in a welded tuff. Most closely spaced in the densely welded zone, a few of the joints are continuous into the partially welded zones above and below, but generally do not extend downward into the nonwelded zone. Joints generally grow parallel to the cooling gradient of the tuff and thus are perpendicular to cooling surfaces. Most commonly vertical, joints may grow at moderate angles from vertical near buried topography. Where vertical heat flow is concentrated along fumarolic pipes within the tuff, joints tend to radiate outward, forming a plumose pattern. Secondary mineralization associated with fumarolic activity commonly makes tuffs more resistant to erosion so that fossil fumaroles are manifested as mounds on the surface of tuffs.

$$T(z) = T_0 + \left(T_s - T_0\right)\left(\frac{z}{z_s}\right),$$

for the upper portion of the tuff, where T is temperature (subscripts s and 0 denote solidus and ambient temperatures, respectively), z is depth, and z_s is the depth of the brittle-ductile transition. The simple, conductive, model-dependent stress distribution (vertical stress neglected and horizontal stresses equal) shows a variation dependent upon depth:

$$\sigma_x = \sigma_y = \frac{\alpha E\left(T_s - T_0\right)}{1-v}\left(1 - \frac{z}{z_s}\right),$$

where horizontal stresses are denoted by σ_x and σ_y, α is the thermal expansion coefficient, E is Young's modulus, and v is Poisson's ratio. Because the horizontal stresses are positive, the tuff is in tension. This cooling and stress magnitude structure is illustrated in Figure 2.3.6.

For the elastic chilled region, deformation leads to stress buildup and concentration at discontinuities (e.g., lithic fragments) where fractures nucleate. Fractures tend to elongate incrementally when stress concentration at a fracture tip (denoted by K_I, which is the stress intensity factor) increases above fracture toughness (K_c). With each incremental fracture growth, stress concentration is temporally relieved only to build up again with further cooling.

For a joint that is uniformly loaded, the stress intensity factor is commonly expressed (Lawn and Wilshaw, 1975) as:

$$K_I = 1.2\sigma_x \sqrt{\pi c},$$

where c is the fracture length. The fracture toughness is typically in the range of 1–4 MPa m$^{1/2}$. This incremental growth model is illustrated in Figure 2.3.7.

This model of joint growth predicts a growth increment dependent upon joint spacing. For the case when a is equal to the joint half length, and b is the horizontal half-spacing, Figure 2.3.8 shows the relationship of growth increment (c/a) to normalized joint spacing as a function of the overall reduction in the fracture tip stress concentration. Overall, when applied to a tuff overlying a nearly horizontal substrate, this model predicts vertically oriented joints (hence fractures) with spacing that tends to decrease

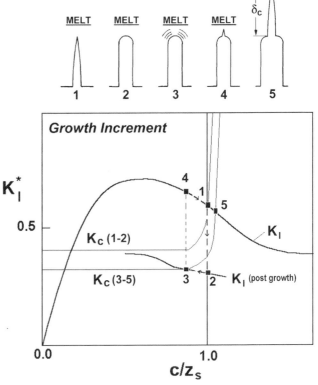

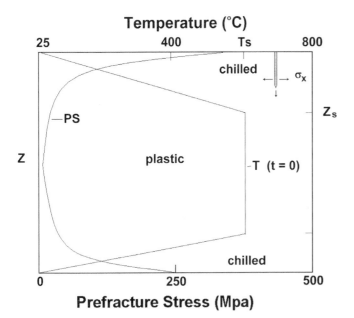

Figure 2.3.6. A plot adapted from DeGraff and Aydin (1993) showing the thermal profile (T) at some initial time ($t = 0$) and the prefracture stress magnitude as a function of depth (z). In this model, horizontal stresses (σ_x) are greater than vertical stress, the bottom and top portions of the tuff are chilled and brittle, but the central portion (below a depth of z_s) is still viscoplastic.

Figure 2.3.7. Plot adapted from DeGraff and Aydin (1993) showing five stages of the incremental growth of a columnar joint fracture into cooling tuff. K_I^* is the normalized stress intensity factor, and K_c is thermally dependent fracture toughness. K_I^* is shown as a function of c/z_s, the ratio of fracture length to depth of the solidus isotherm, which decreases constantly with cooling but increases abruptly with incremental growth. With deformation near or within viscous tuff (1) ($T \geq T_s$), the fracture tip becomes blunted (2), then cooling causes the solidus surface to move away from the blunted fracture, causing stress to build up (3) and eventually concentrate into a sharp tip (4), where K_I^* is greater than K_c, leading to incremental growth δ_c (5).

as cooling rate increases. Such a prediction indicates that thicker welded tuffs will tend to have more widely spaced joints than thinner tuffs, if cooling is dominantly conductive.

FRACTURE CHARACTERISTICS

Detailed fracture investigations of the upper member of the Bandelier Tuff of northern New Mexico (Vaniman and Wohletz, 1990; Kolbe et al., 1995; Vaniman and Chipera, 1995; Walters, 1996; Reneau and Vaniman, 1998; Wohletz, 1995a, 1995b, 1996, 1998) provide data that illustrate the distribution and character of fractures in welded tuff. Figure 2.3.9 shows a geological map of the central portion of Los Alamos National Laboratory, which is situated on the Pajarito Plateau. The upper member of the Bandelier Tuff is subdivided into four units of varying degrees of welding (Fig. 2.3.10). Incision by west-to-east–running canyons exposes cliff sections of the tuff, which has allowed detailed fracture mapping. Seven horizontal transects have been

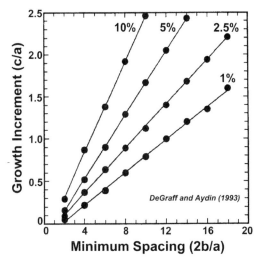

Figure 2.3.8. From DeGraff and Aydin (1993), this plot shows that where joints are more widely spaced with respect to their half length, the growth increment will be larger. Solutions are shown for a range of overall joint tip stress intensity decrease due to growth. Where more stress is relieved during growth, the growth increment will be larger.

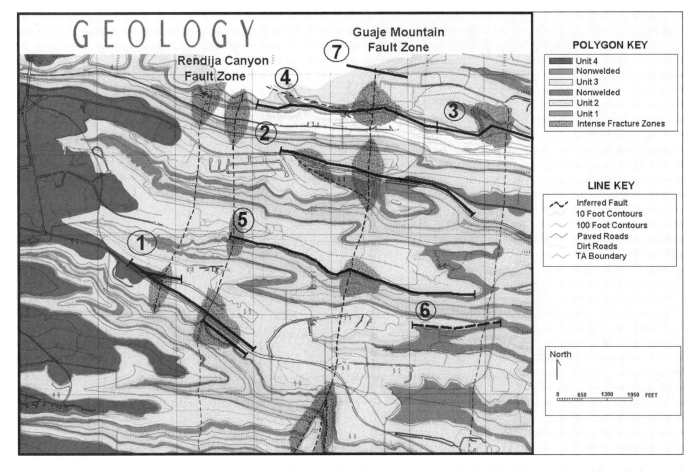

Figure 2.3.9. Geological map of the central part of Los Alamos National Laboratory area on the Pajarito Plateau (adapted from Vaniman and Wohletz, 1993). Bandelier Tuff units are shown in blues, greens, yellows, and red. Brown stippled areas are zones of intense fracturing, mostly along mapped traces of the Rendija Canyon fault and Guaje Mountain fault. Seven fracture transects are shown: (1) along the north side of Two Mile Canyon; (2) along the north side of Sandia Canyon; (3) lower Los Alamos Canyon; (4) upper Los Alamos Canyon; (5) Mortandad Canyon, (6) the north side of Acid Canyon; and (7) just north of Los Alamos Canyon.

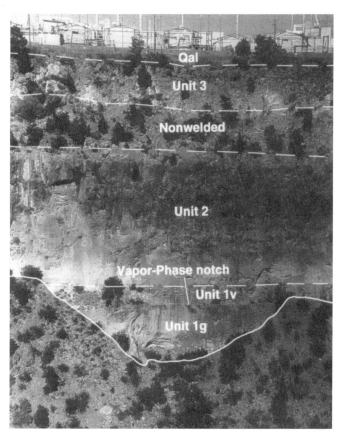

Figure 2.3.10. Photograph of the northern cliff face of Los Alamos Canyon near DP Mesa. From bottom to top, the stratigraphy of the tuff includes nonwelded Unit 1, partially covered by Quaternary talus and consisting of glassy and vapor-phase altered portions. Unit 2 is moderately welded and overlain by a nonwelded transition into Unit 3 (partially welded), which is in turn overlain by Quaternary alluvium. Note that Unit 4 has been presumably removed by erosion in this area.

ture strikes are apparently controlled by local stress patterns that existed on the Pajarito Plateau during and after tuff emplacement, fracture dips are generally vertical, fracture apertures average ~1 cm, and fracture fillings exist mainly within 15 m of the surface and are derived from infiltrated clay and calcite derived from soil. As discussed later in this paper, cooling fracture characteristics are intimately tied to tectonic displacement that occurred both before and after the tuff emplacement, cooling, and compaction.

Fracture Density

Fracture density was calculated as the number of fractures in 100 ft intervals centered on each fracture. Average fracture spacing was found to be fairly constant over all areas of investigation at ~5 ft (20/100 ft) in studied areas, but over tectonic zones, the fracture density exceeded values of >60/100 ft (Fig. 2.3.12). This marked increase of fracture linear density over tectonic zones is not readily apparent from outcrop observations. As shown in Figure 2.3.12, the density variation expressed as number of fractures in 10 ft intervals centered on each fracture is small, generally unrecognizable to the eye and only readily apparent from data plots representing density over 100 ft intervals. However, in regions outside studied areas where the thickness of welded units is about one-third of that for the measured transects, average fracture spacing does decrease to ~2 ft, which supports the predictions of the joint formation model.

Fracture Strike

In general, variability in fracture strike causes vertical fractures to intersect and form polygonal columns. Many fractures are sinuous, showing curvature in strike. Fracture strikes show a crude polymodality in distribution (Table 2.3.2), which for the E-W profiles documented can be characterized as NW- and NE-trending fracture sets, but an E-W set is theoretically extant (Fig. 2.3.13). Because N-S exposure is extremely limited, as is mesa-top fracture exposure, it is difficult to prove the calculated distribution shown in Figure 2.3.13. The calculations do show a fracture distribution that produces 60° intersections, which perhaps fits tensional fracture of an isotropic brittle material better. Because most fractures are either part of the NW- or NE-trending sets, fracture intersection produce 4-sided columns that have orthogonal to rhombic angles. Fractures tend to be orthogonal over fault zones and are slightly rotated eastward (Fig. 2.3.14). Figure 2.3.15 is a three-dimensional projection of fractures measured along an E-W section. Although the calculated N-S cross section does not show a marked dominance of E-W–trending fractures, it qualitatively shows that where fractures are more closely spaced, they tend to intersect at lower angles—the larger fracture-bounded blocks tend to be more orthogonal.

Fracture Dip

Most fractures tend to be nearly vertical, although more horizontally oriented ones tend to form along planes that mark changes

investigated, generally running along the northern slopes of canyons where fractures are best exposed. Because these mapped transects cross two major fault zones (Rendija Canyon fault zone and Guaje Mountain fault zone), the investigations revealed the relationship of cooling joint fractures to fractures possibly caused by tectonic movement.

The method of investigation involved creation of photomosaics along the canyon walls to serve as base maps for fracture mapping (Fig. 2.3.11). The photographs were taken so that most fractures could be readily identified. Each fracture was assigned a number, and for each, several characteristics were systematically measured and added to a database for later analysis. The best exposures of fractures were noted in Unit 2 and Unit 3, and most fractures could be traced through both units. Because the investigation methods applied to environmental efforts using engineering plans with distance measured in feet, the distance unit is retained in the following sections.

The results of this study, summarized in the following sections, show that fracture density (spacing) averages 20/100 ft, frac-

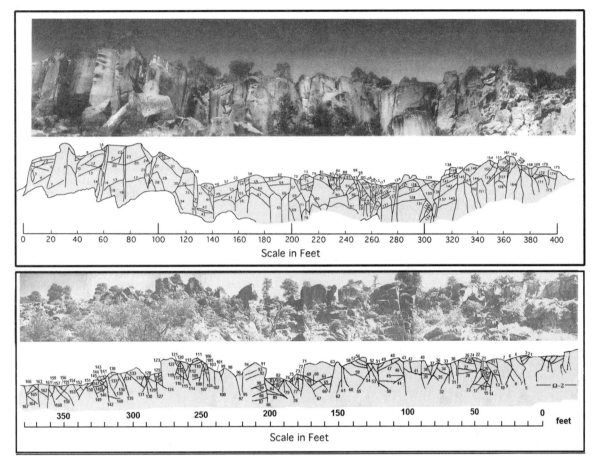

Figure 2.3.11. Examples of two fracture maps for Los Alamos Canyon, the top one depicting an area where fractures are well exposed by a cliff face, and the bottom one, a region where fractures are poorly exposed.

in tuff properties (e.g., welding). Most fractures are sinuous in dip, but a few are very planar. Table 2.3.3 shows that the average dip is from 70° to 80°, and that there is little variation of dip between strike sets (NW and NE), between background fractures and those in a fault zone, and between northerly and southerly dipping fractures. An observation true for the Bandelier Tuff is that measured dips along canyon walls tend to be in a direction away from the canyon walls. This observation may stem from the likelihood that fractures dipping toward the canyon are more likely to spall during sidewall erosion and thus are not preserved. Figure 2.3.16 shows variation of fracture dip angle for all fractures, northerly, and southerly dipping fracture sets. Note that over the fault zone, fractures become less steep, and the angle between northerly and southerly dipping fractures tends to grow large. However, from Table 2.3.3, it is apparent that the fault zone is not readily apparent from dip variation on the NE and NW fracture sets.

Aperture

Fractures show variable aperture because of fracture sinuosity, as described above in fracture dip characteristics explanation.

TABLE 2.3.2. FRACTURE STRIKE DATA

Fracture Set	Number	Mean Strike	Standard deviation
All Fractures	4940	N9E	±46
NE	2820	N44E	±25
NW	2120	N36W	±25
Background (west)	1085	N5E	±46
NE	557	N44E	±26
NW	528	N35W	±22
Fault zone	933	N10E	±50
NE	527	N47E	±24
NW	406	N39W	±25
Background (east)	1297	N13E	±48
NE	756	N43E	±24
NW	541	N30W	±22

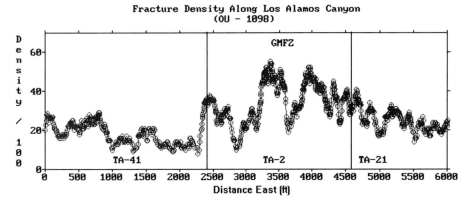

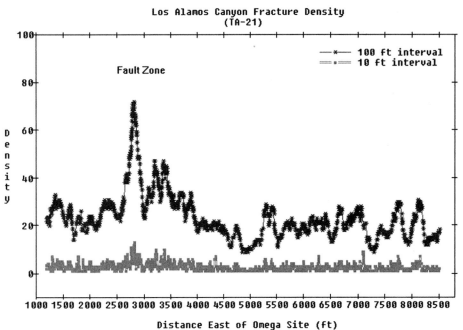

Figure 2.3.12. Two examples of fracture density measurements. The top plot displays fracture density that increases near Omega site where the Guaje Mountain fault zone extends. The bottom plot shows measurements of densities under Material Disposal Area V (MDA-V). The bottom curve in the lower plot shows density calculated as the number of fractures per 10 ft interval, which is more what a field observer would perceive, perhaps masking the larger-scale trends.

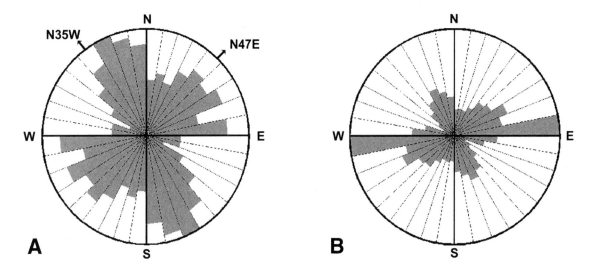

Figure 2.3.13. Fracture strike distributions. (A) Measured (apparent) distribution showing two modes characterizing NW- and NE-trending fracture sets. Because fractures were observed along a generally E-W transect, fractures parallel to this transect are not well exposed. By calculating the effect of geometry on exposure, a hypothetical distribution (B) shows that in fact E-W–trending fractures may dominate.

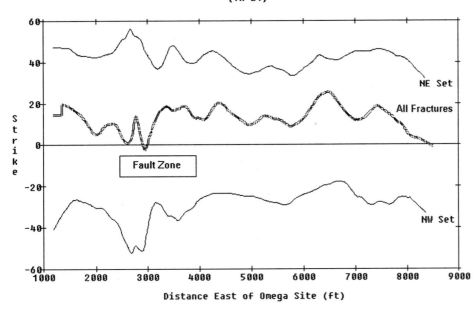

Los Alamos Canyon Fracture Strikes
(TA-21)

Distance East of Omega Site (ft)

Figure 2.3.14. Variation of fracture strike from W to E along a portion of Los Alamos Canyon. This plot shows average krieged values for all fractures and for the NW and NE sets individually. Note that the angle between fracture sets remains fairly constant at ~60°, except over the fault (fracture) zone, where the average value varies antithetically. Figure 2.3.14. Variation of fracture strike from W to E along a portion of Los Alamos Canyon. This plot shows average krieged values for all fractures and for the NW and NE sets individually. Note that the angle between fracture sets remains fairly constant at ~60°, except over the fault (fracture) zone, where the average value varies antithetically.

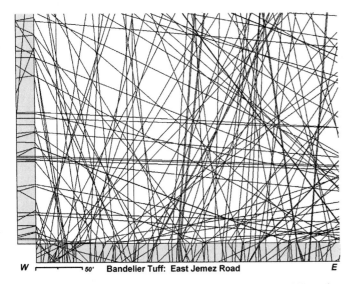

Figure 2.3.15. Projection of fractures measured on an E-W section showing a more acute (rhombic) intersection where fractures are more closely spaced.

TABLE 2.3.3. FRACTURE DIP DATA

Fracture set	Number	Mean dip from vertical (°)	Standard deviation
All fractures	4940	78N	±26
NE	2820	77N	±26
N	1447	69N	±26
S	393	67S	±20
NW	2120	81N	±25
N	1165	71N	±23
S	310	60S	±23
Background (west)	1085	81N	±29
NE	557	77N	±32
NW	528	84N	±27
Fault zone	933	78N	±34
NE	527	76N	±33
NW	406	79N	±36
Background (east)	1297	82N	±26
NE	756	81N	±26
NW	541	83N	±26

TABLE 2.3.4. FRACTURE APERTURE DATA

Fracture set	Number	Mean aperture (cm)	Standard deviation
All Fractures	4940	0.95	±1.73
NE	2820	0.98	±1.86
NW	2120	0.91	±1.46
Background (west)	1085	0.65	±0.92
NE	557	0.69	±0.97
NW	528	0.62	±0.85
Fault zone	933	0.97	±1.54
NE	527	0.94	±1.37
NW	406	1.18	±1.76
Background (east)	1297	0.69	±0.74
NE	756	0.69	±0.77
NW	541	0.67	±0.72

Overall, fracture average apertures range from 0.7 to 1.0 cm, but average fracture aperture in the tectonic zone increases to over 2.0 cm (Table 2.3.4).

Figure 2.3.17 shows variation in fracture aperture along Los Alamos Canyon. Over nontectonic regions, the average aperture fluctuates between 0.2 and 1.0 cm, but over fault zones, it can increase to over 3.0 cm where fracture density is also much higher. In general, fracture aperture increases with fracture dip such that nearly horizontal fractures are generally closed and vertical ones have well-developed apertures (Fig. 2.3.18), a relationship that supports the model that tuff cooling fractures accommodate horizontal strain.

Fracture Fill

Apparently physically continuous with soils on the tuff surface, fracture fill materials exist downward 3–15 m below the surface, below which fractures are open. These materials are dominantly composed of illuviated clays (smectite, illite, and kaolinite) derived from weathering of the tuffaceous soils that infiltrated open fractures, hence they are pedogenic. Precipitated calcite, derived from soil water, fills in pores and shrinkage cracks in clays, forming vertical laminae common in fracture fill materials where apertures are greater than several centimeters. The fill material also includes minor amounts of mineral grains and fragments from the tuff. A crude zonation of fill materials is evident in some places where weathered tuff and clay minerals are most common near the fracture faces and infiltrated soil detritus and calcite is more abundant in central portions of the filled zone (Figs. 2.3.19 and 2.3.20). Because of the pedogenic origin of fracture fill materials, fractures have been subjected to oxidizing effects of water infiltration. Oxidation alteration is manifested on fracture faces and tuff immediately adjacent to fractures, where a zone of alteration up to several centimeters thick is marked by iron-oxide staining (Fig. 2.3.21).

The Role of Cooling Fractures in Tectonic Displacement

Overall the character of fractures in the Bandelier Tuff indicates that their ultimate origin is tied to columnar jointing by cooling contraction of the tuff. However, tectonics may likely have played an important role in determining the substrate over which the tuff compacted, and, subsequently, by imparting stress upon the tuff that would ultimately be relieved by opening of these cooling fractures. This conclusion is supported by the increases

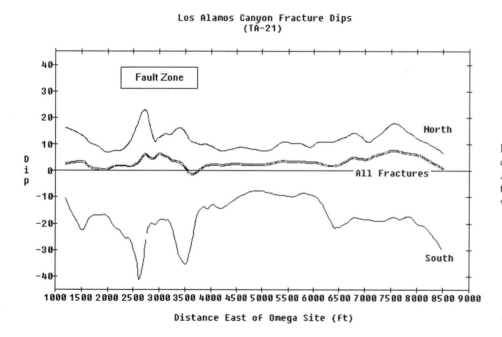

Figure 2.3.16. E-W variation of fracture dips (measured from vertical) along Los Alamos Canyon. Considerable fluctuation of dips occurs over the fault zone, where dips tend to be less vertical.

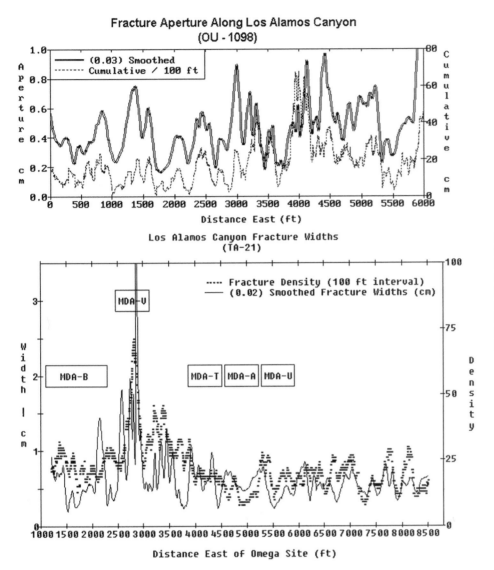

Figure 2.3.17. The top plot shows fracture aperture variation over a nontectonic region near Omega site in Los Alamos Canyon (3500 ft east). The bottom plot shows this variation further to the east over a branch of the Guaje Mountain fault (near MDA-V). Note in the lower part how fracture aperture mimics fracture density variation.

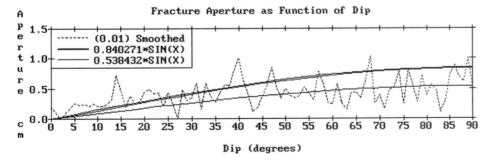

Figure 2.3.18. Variation of fracture aperture with fracture dip. A sinusoidal function adequately expresses this variation.

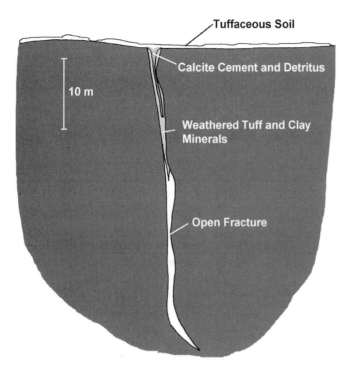

Figure 2.3.19. Schematic illustration of fracture fill materials for the Bandelier Tuff. Note the crude zonation of fill materials where weathered tuff and clay minerals dominate the fill near the fracture faces, and infiltrated calcite cement and soil detritus is more abundant in the central portions of the filling material.

of fracture linear density and average fracture aperture as well as the rotation of fracture orientation noted over fault zones.

Figure 2.3.22 is a schematic illustration of the role of tectonics in fracture orientation. When the tuff was emplaced, it ponded and was thicker in the canyons. During its cooling and compaction, the greatest vertical displacement of the tuff surface occurred in these canyons, concentrating tensional stresses along the canyon walls with cooling fractures developing parallel to the canyons. Another effect of tectonics was regional extensional stress (Aldrich et al., 1986), which influenced the orientation of cooling fractures as the tuff contracted during its cooling. This horizontal stress orientation resulted in an antithetic network of cooling fractures, separated by an angle of 60°–90°, centered on the direction perpendicular to the regional extension. The role of tectonic stresses is detailed by Walters (1996).

As noted already, traces of the Rendija Canyon and Guaje Mountain faults are mapped across the Pajarito Plateau. North of Los Alamos, these faults display from several to tens of meters of vertical offset in surface rocks, but south of Los Alamos Canyon, the surface offset is difficult to measure. There is some indication of the offset by changes in surface slope over the faults, but little, if any, notable vertical offset of the tuff. Hypotheses regarding these observations include that either the fault offset decreases to the south to be negligible in the area of fracture studies or that the tuff has acted as a concealing unit and has accommodated fault offset by incremental movement on cooling fractures spread out over a wide region above the

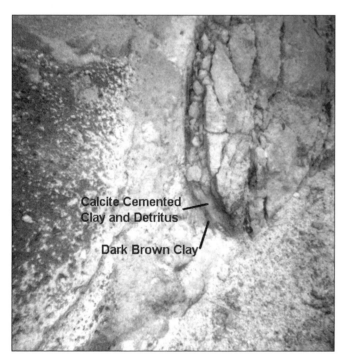

Figure 2.3.20. Close-up photograph showing the crude zonation of fracture fill materials. The calcite-cemented clay and detritus is a lighter color than the clay filling near the face of the fracture.

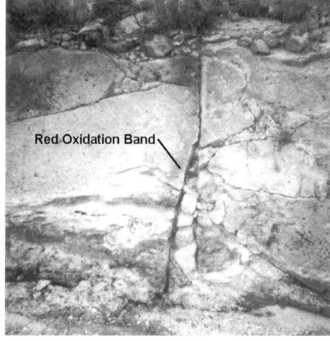

Figure 2.3.21. Reddish iron-oxidation stain along a fracture filled with 1–2 cm of clay and detritus; this fracture is only several feet below the tuff surface.

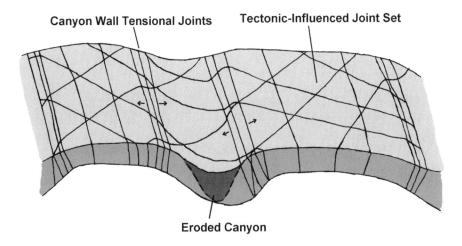

Canyon Wall Tensional Joints   **Tectonic-Influenced Joint Set**

Eroded Canyon

Figure 2.3.22. Schematic illustration of the effects of regional tectonism on the location and distribution of cooling fractures in a welded tuff.

fault. As demonstrated by the theoretical modeling and fracture characteristics noted herein, tuff cooling fractures accommodate horizontal strain by their aperture. By simple trigonometric projection, fracture aperture can be shown to accommodate vertical strain as well (Fig. 2.3.23).

By assuming that fracture aperture has developed by vertical displacement of the tuff in response to fault movement, a simple algorithm can be applied to all fractures that utilizes the fracture strike, dip, and aperture. Fractures striking perpendicular or nearly perpendicular to a fault trace do not produce any vertical displacement across a fault. However, fractures striking along or at some acute angle to a fault trace can produce vertical offset. Fractures dipping to the west produce down-drop to the east, whereas fractures dipping easterly produce down-drop to the west. The amount of down-drop is a function of fracture dip angle and aperture. By applying this trigonometric algorithm to a fracture profile that crosses a fault zone, one can see the potential vertical offset the fault has produced.

Figure 2.3.24 shows an example for a portion of Los Alamos Canyon cut by the trace of the Guaje Mountain fault. Hypothetical fracture vertical displacement is shown as sum of individual fracture displacement (negative values are west-side-down, and positive values are east-side-down) for 100 ft intervals centered on each fracture. By smoothing these data, fracture displacement is negligible (<10 cm) along most of the profile except over the fault zone, which extends ~1000 ft, from 3400 to 4400 ft east. Because this fault is known to have produced down-throw to the west, one can view the cumulative vertical displacement by sequentially summing the displacements on each fracture from east to west. Figure 2.3.25 shows that over the Guaje Mountain fault zone, a total cumulative displacement of ~3 m down to the west occurs over a distance of 700 ft. West of this zone, the cumulative displacement changes little. Figure 2.3.26 shows the same kind of plot as that in Figure 2.2.25, except it is for a profile along Mortandad Canyon, where both the Rendija Canyon and Guaje Mountain fault traces project. In Figure 2.3.26, the western shoulder of the Guaje Mountain fault shows

nearly the same profile in Mortandad Canyon as is documented for Los Alamos Canyon in Figure 2.3.25.

DISCUSSION

Overall, cooling fracture characteristics appear to have a relationship to tectonic stresses and strains. When viewing the data sets for these fractures, the distribution of characteristics does not appear to have any discernible patterns—that is the strike, dip, and aperture of one fracture looks to be independent from that of the next fracture, and so on. However, after applying a trigonometric algorithm that considers strike, dip, and aperture, a pattern along a fracture traverse becomes apparent. If one assumes that this pattern does in fact reflect the cumulative vertical displacement along fractures, then one sees that fracture characteristics must not be randomly distributed, but can be related to tectonism, and furthermore, the pattern produced by this trigonometric relationship is reproducible at different locations. To be sure, this trigonometric pattern may be due to some other nontectonic process, but as of yet that explanation has not been explored.

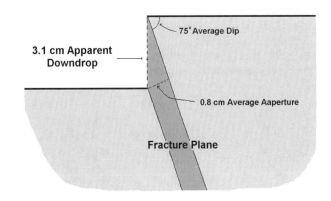

75° Average Dip

3.1 cm Apparent Downdrop

0.8 cm Average Aaperture

Fracture Plane

Figure 2.3.23. Schematic illustration of a dipping fracture with aperture produced by vertical displacement.

Calculated Vertical Displacement on Fractures

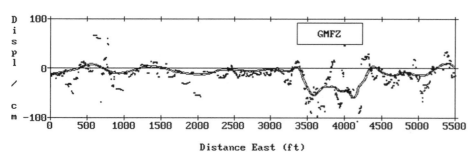

Figure 2.3.24. Plot of cumulative apparent fracture vertical displacement per 100 ft interval centered on each fracture. Note that cumulative displacement is nearly zero except over the Guaje Mountain fault zone, where it is negative (down to the west).

- 100-ft cumulative displacement
— (0.1) Smoothed 100-ft cumulative displacement

Calculated Cumulative Vertical Displacement on Fractures

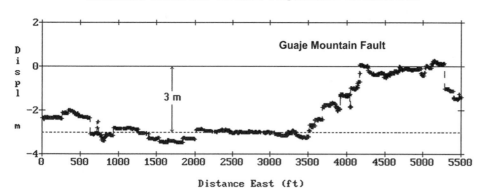

Figure 2.3.25. Cumulative apparent vertical displacement on fractures (summed from east to west), showing a 3 m downdrop over the Guaje Mountain fault zone (GMFZ).

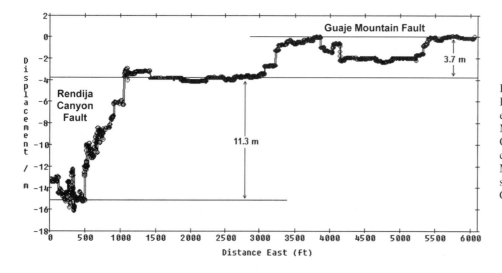

Figure 2.3.26. Similar to the plot in Figure 2.3.25, this cumulative apparent fracture displacement profile is for Mortandad Canyon where both Rendija Canyon and Guaje Mountain fault traces cross. The west shoulder of the Guaje Mountain fault shows a very similar offset and profile as it does in Los Alamos Canyon, depicted in Figure 2.3.25.

One may ask why cooling fractures are more abundant in fault zones if tectonic movement did not occur during the relatively short span of time during which cooling contraction occurred. The first answer to this question can be given for the circumstance that tectonism occurred prior to the tuff emplacement and cooling. In that case, there is a likelihood that the tuff was emplaced over a fault scarp, and differential compaction over the fault scarp might have concentrated stresses, requiring more fractures to incrementally accommodate the strain. A second answer lies in the fact that not all cooling fractures are equally well developed, such that many might have developed only a plane of minute dislocation or weakness without visible breakage, and some may have been reannealed during the welding process. In such circumstances where tectonic displacement occurs after cooling contraction, the added stress buildup in the tuff during faulting may cause these incipient or reannealed fractures to open up and become visible.

CONCLUSIONS

Welded tuffs develop fractures (joints) during cooling contraction that occurs after emplacement of the tuff. Best developed in welded zones, these fractures can extend throughout the entire thickness of tuff. Fracture spacing is on the order of meters and is smallest where cooling occurred rapidly and/or where the tuff has been faulted. Though dominantly vertically oriented, horizontal and plumose joint sets do form where cooling is influenced by substrate irregularities and fumarolic processes. It is typical to find fractures with apertures up to 1 cm or more. These apertures may be filled by clay minerals and detritus, especially in locations within several meters of the surface.

Fracture characteristics of strike, dip, and aperture may appear to be randomly distributed from field observations; however, there is a possibility that a stochastic relationship between these characteristics exists because of preexisting topographic fabric and postemplacement tectonic stresses.

The importance of tuff fractures is greatest in areas where a population center and associated industries exploit it for building material, aquifers, and waste disposal. Under saturated conditions, fractures promote infiltration of meteoric water and contaminants that may be placed at the surface or within the tuff, leading to widespread contaminant dispersal. Fractured tuffs can also display considerable slope instability near canyon walls by mass wasting associated with block falls. On the other hand, fractured tuffs display features that make them good aquifers where they are below the water table. In this setting, they are good hosts for geothermal reservoirs in areas of high heat flow. Also, saturated conditions may lead to rapid alteration of tuffs, such that fractures become sealed by zeolites and other secondary minerals. In this situation, fractured tuffs become aquitards.

ACKNOWLEDGMENTS

This work was done under the auspices of the U.S. Department of Energy. The author especially recognizes Michael F. Sheridan for his pioneering work on mineralogy, petrology, and compaction of welded tuff; work that stimulated and guided much of efforts described in this manuscript. Grant Heiken promoted the publication of this work and provided numerous editorial improvements. Both have been great teachers and good friends.

REFERENCES CITED

Aldrich, M.J., Jr., Chapin, C.E., and Laughlin, A.W., 1986, Stress history and tectonic development of the Rio Grande Rift, New Mexico: Journal of Geophysical Research, v. 91, p. 6199–6211.
Barton, C.C., and Larsen, E., 1985, Fractal geometry of two-dimensional fracture networks at Yucca Mountain, southwestern Nevada, *in* Proceedings of the International Symposium on Fundamentals of rock joints: Bjorkliden, Sweden, International Symposium Proceedings, p. 77–84.
DeGraff, J.M., and Aydin, A., 1993, Effect of thermal regime on growth increment and spacing of contraction joints in basaltic lava: Journal of Geophysical Research, v. 98, p. 6411–6430.
Fisher, R.V., and Schmincke, H.-U., 1984, Pyroclastic rocks: Berlin, Springer-Verlag, 472 p.
Francis, P., 1993, Volcanoes a planetary perspective: New York, Oxford University Press, Inc., 443 p.
Fuller, C.M., and Sharp, J.M., Jr., 1992, Permeability and fracture patterns in extrusive volcanic rocks: Implications from the welded Santana Tuff, Trans-Pecos, Texas: Geological Society of America Bulletin, v. 104, p. 1485–1496, doi: 10.1130/0016-7606(1992)104<1485:PAFPIE>2.3.CO;2.
Kolbe, T., Sawyer, J., Springer, J., Olig, S., Reneau, S., Hemphill-Haley, M., and Wong, I., 1995, Evaluation of the potential for surface faulting at TA-63, draft final report: Oakland, California, Woodword-Clyde Federal Services, unpublished report.
Lawn, B.R., and Wilshaw, T.R., 1975, Fracture of brittle solids: Cambridge, Cambridge University Press, 204 p.
Reneau, S.L., and Vaniman, D.T., 1998, Fracture characteristics in a disposal pit on Mesita del Buey: Los Alamos, Los Alamos National Laboratory Report LA-13539-MS, 21 p.
Ross, C.S., and Smith, R.L., 1961, Ash-flow tuffs: Their origin, geologic relations, and identification: U.S. Geological Survey Professional Paper 366, 81 p.
Vaniman, D., and Chipera, S., 1995, Mesa-penetrating fractures, fracture mineralogy, and projected fault traces at Pajarito mesa, *in* Geological site characterization for the proposed mixed waste disposal facility, Los Alamos National Laboratory: Los Alamos, Los Alamos National Laboratory Report LA-13089-MS, p. 71–86.
Vaniman, D., and Wohletz, K., 1990, Results of geological mapping/fracture studies, TA-55 area: Los Alamos, Los Alamos National Laboratory Seismic Hazards Memo EES1-SH90-17, 25 p.
Vaniman, D., and Wohletz, K., 1993, Reconnaissance geology of north-central LANL: Los Alamos, New Mexico, Los Alamos National Laboratory, Facility for Information Management, Analysis, and Display Map G101599, scale 1:7800.
Walters, M.C., 1996, Fracture analysis of the Bandelier Tuff, Pajarito Plateau, north-central Rio Grande Rift, New Mexico [MSc. thesis]: Fort Worth, Texas, Texas Christian University, 91 p.
Wohletz, K.H., 1995a, Measurement and analysis of rock fractures in the Tshirege Member of the Bandelier Tuff along Los Alamos Canyon adjacent to TA-21, *in* Broxton, D.E., and Eller, P.G., eds., Earth science investigations for environmental restoration—Los Alamos National Laboratory Technical Area 21: Los Alamos, Los Alamos National Laboratory report LA-12934-MS, p. 19–31.
Wohletz, K.H., 1995b, Fracture characterization of the DP Tank Farm: Los Alamos National Laboratory Environmental Restoration Technical Communication, Los Alamos National Laboratory, August 1995, 13 p.
Wohletz, K.H., 1996, Fracture characterization of the Bandelier Tuff in OU-1098 (TA-2 and TA-41): Los Alamos, Los Alamos National Laboratory Report LA-13194-MS, 19 p.
Wohletz, K.H., 1998, Fracture characterization along a portion of Mortandad Canyon: Los Alamos National Laboratory Environmental Restoration Technical Communication, Los Alamos National Laboratory, September 1998, 22 p.

MANUSCRIPT ACCEPTED BY THE SOCIETY 29 DECEMBER 2005

Geological Society of America
Special Paper 408
2006

Chapter 3

Geotechnical properties of tuffs at Yucca Mountain, Nevada

compiled by

Mala Ciancia
URS Corporation, 201 Willowbrook Boulevard, Wayne, New Jersey 07474-0290, USA

edited by

Grant Heiken
331 Windantide Place, Freeland, Washington 98249-9683, USA

INTRODUCTION

Geotechnical properties of the tuffs at Yucca Mountain include the physical, mechanical, thermal, thermal/mechanical, and other relevant special properties.

Stratigraphic Framework for Testing

Rocks that are important to Yucca Mountain repository design are for the most part within the Miocene-age Paintbrush Group, which is composed of welded and nonwelded ignimbrites. The formations within the Paintbrush Group, as shown in Table 3-1 include, in ascending order, the Topopah Spring Tuff, the Pah Canyon Tuff, the Yucca Mountain Tuff, and the Tiva Canyon Tuff. Bedded tuffs are found separating all of the ignimbrite units of the Paintbrush Group. These tuffs are typically nonwelded and are nonlithified to moderately lithified. They range from 0 to 10 m in thickness and contain a variety of ignimbrite and ash-fall

---

Editor's Note: Yucca Mountain, Nevada has been the object of intense geologic study for the past 20 years because it was chosen as a candidate site for long-term storage of high-level nuclear waste in the United States. Yucca Mountain is made up of a thick sequence of Miocene-age ignimbrites and ash-fall tuffs that were erupted during the formation of the Timber Mountain caldera complex. Because of the high standards required for characterizing this site, Yucca Mountain is the most intensely studied group of tuffs in the world. The large team responsible for geoengineering has graciously allowed us to use the excellent and comprehensive data set collected at Yucca Mountain, which is summarized in this chapter. The data sets presented in this chapter are mostly from tuff units that enclose the proposed repository and were collected from drill core and in situ tunnel measurements.

deposits, and thin tuffaceous sandstones. Five of these bedded tuff sections have been recognized within the Paintbrush Group (Brechtel et al., 1995).

Below the rocks of the Paintbrush Group is the Calico Hills Formation. Overlying the Paintbrush Group in local areas near Exile Hill are younger, nonwelded ignimbrite and ash-fall tuffs, including tuff unit "X" and the Rainier Mesa and Ammonia Tanks Tuffs of the Timber Mountain Group.

A thermal/mechanical stratigraphy was developed to provide a systematic basis for characterizing the rock mass in the site based on geoengineering properties of the rock units. The stratigraphy is based on thermal and mechanical rock characteristics that are important to repository design, and was developed by designating lithologic units, in whole or part, or a group of contiguous units, or parts, as thermal/mechanical units. This nomenclature was first proposed by Ortiz et al. (1985) to group rocks with similar thermal and mechanical properties. The stratigraphy was based on the observation (Lappin et al., 1982) that thermal and mechanical properties can be correlated directly to grain density and porosity. The stratigraphy of Ortiz et al. (1985) includes 16 thermal/mechanical units, seven of which are shown on Table 3-1. The thermal/mechanical units were originally identified megascopically in terms of their welding and lithophysal cavity content.

The definition of thermal/mechanical units reflects to a large extent the general degree of welding, and the defined units correlate generally with groups of lithostratigraphic units, or in the case of the Topopah Spring Tuff, parts of a lithostratigraphic unit

Table 3-1. Comparison of Several Stratigraphic Subdivisions of Volcanic Rocks at Yucca Mountain and Encountered on the Yucca Mountain Site Characterization Project

Lithostratigraphic Units[1]		Thermal/Mechanical Units[2]	Hydrogeologic Units[3]
Timber Mountain Tuff (Tm)	Rainier Mesa member (Tmr)	Undifferentiated overburden (UO)	Unconsolidated surficial materials (UO)
	Pre-Rainier Mesa bedded tuff (Tmrbt1)		
Tuff unit "X" (Tpk)	tuff unit "X" (Tpki)		
	post-Tiva Canyon bedded tuff (Tpbt5)		
PAINTBRUSH GROUP (Tp)			
Tiva Canyon Tuff (Tpc)	crystal-rich member (Tpcr) vitric zone (Tpcrv) - nonwelded subzone (Tpcrv3) - moderately welded subzone (Tpcrv2) - densely welded subzone (Tpcrv1) nonlithophysal zone (Tpcrn) lithophysal zone (Tpcrl)	Tiva Canyon welded (TCw)[4]	Tiva Canyon welded (TCw)
	crystal-poor member (Tpcp) upper lithophysal zone (Tpcpul) middle nonlithophysal zone (Tpcpmn) lower lithophysal zone (Tpcpll) lower nonlithophysal zone (Tpcpln) - hackly subzone (Tpcplnh) - columnar subzone (Tpcplnc) vitric zone (Tpcpv) - densely welded subzone (Tpcpvv) - moderately welded subzone (Tpcpvm) - nonwelded subzone (Tpcpvn)	Paintbrush Tuff nonwelded (PTn)	Paintbrush nonwelded (PTn)
	pre-Tiva Canyon bedded tuff (Tpbt4)		
Yucca Mountain Tuff (Tpy)	Yucca Mountain Tuff (Tpyt)		
	pre-Yucca Mountain bedded tuff (Tpbt3)		
Pah Canyon Tuff (Tpp)	Pah Canyon Tuff (Tppt)		
	pre-Pah Canyon bedded tuff (Tpbt2)		
Topopah Spring Tuff (Tpt)	crystal-rich member (Tptr) vitric zone (Tptrv) - nonwelded subzone (Tptrv3) - moderately welded subzone ((Tptrv2) - densely welded subzone (Tptrv1) nonlithophysal zone (Tptrn) lithophysal zone (Tptrl)	Topopah Spring welded, lithophysae-rich (TSw1)	Topopah Spring welded (TSw)
	crystal-poor member (Tptp) upper lithophysal zone (Tptpul) [upper pt]		
REPOSITORY HOST HORIZON	upper lithophysal zone (Tptpul) [lower pt] middle nonlithophysal zone (Tptpmn) lower lithophysal zone (Tptpll) lower nonlithophysal zone (Tptpln)	Topopah Spring welded, lithophysae-poor (TSw2)	
	vitric zone (Tptpv) -densely welded subzone (Tptpv3) -moderately welded subzone (Tptpv2) -nonwelded subzone (Tptpv1)	Topopah Spring welded, vitrophyre (TSw3)	Topopah Spring basal vitrophyre (TSbv)
	pre-Topopah Spring bedded tuff (Tpbt1)	Calico Hills nonwelded (CHn)	Calico Hills nonwelded (CHn)
Calico Hills (Tac)	Calico Hills Formation (Tac)		
	pre-Calico Hills bedded tuff (Tacbt1)		

1) Buesch, Spengler et al. 1996a
2) Ortiz et al. 1985
3) Arnold et al. 1995
4) Where preserved, the base of the densely welded subzone forms the base of the TCw thermal-mechanical and hydrogeologic units. (Buesch, Spengler et al. 1996a)

(Table 3-1). The upper two Topopah Spring Tuff units, TSw1 and TSw2, are both within the densely welded, devitrified Topopah Spring Tuff. Originally, Ortiz et al. (1985) defined TSw1 as containing >10% void space from lithophysal cavities and TSw2 as containing <10% void space from lithophysal cavities. (Lithophysae are spherical structures composed of radially oriented alkali feldspar and silica minerals; they are found in glassy silicic lavas and densely welded tuffs. Some lithophysae are hollow). This difference in lithophysae content was considered for repository design, and the top of the proposed repository host horizon is, in fact, aligned with the horizon below which lithophysae content is <10%, as inferred from borehole geophysical data (CRWMS M&O, 1997d). This may be 30–35 m above the Tptpul-Tptpmn (upper lithophysal zone of the crystal-poor member of the Topopah Spring Tuff; middle nonlithophysal zone of the crystal-poor member of the Topopah Spring Tuff) lithostratigraphic contact within the potential emplacement area (Spengler and Fox, 1989). However, the change in percentage of lithophysae does not occur at a consistent stratigraphic position in the crystal-poor upper lithophysal zone, and it can be difficult to identify. For this reason, the Tptpul-Tptpmn lithostratigraphic contact is thus now considered nominally equivalent to and used to define the TSw1-TSw2 thermal/mechanical contact.

Two thermal/mechanical units, the TCw and the TSw1, were further subdivided into the high-lithophysal and non-lithophysal zones for Exploratory Studies Facility geotechnical studies (Brechtel et al., 1995; Kicker et al., 1996), in order to investigate differences in rock structure and rock mass quality within these zones (Brechtel et al., 1995). The zones represent summations of the individual lithophysal and nonlithophysal lithostratigraphic zones for those units. All data presented for the TSw2 thermal/mechanical unit were from the Tptpmn lithostratigraphic unit, and so were not further subdivided.

Most of the Tiva Canyon Tuff is contained in the TCw thermal/mechanical unit. This unit includes rock between and including the densely welded subzone of the vitric zone of the crystal-rich member, and, where it is present, the densely welded subzone of the vitric zone of the crystal-poor member. Where the crystal-poor, vitric, densely welded subzone of the Tiva Canyon Tuff does not occur, the contact of the devitrified rocks in the columnar subzone and the vitric rocks in the moderately welded subzone correspond to the base of the TCw (Buesch, et al., 1996a). The unit is exposed on top of Yucca Crest and the ridges on the eastern flank.

Below the TCw is the PTn thermal/mechanical unit. This unit consists of partially welded to nonwelded, vitric, and, in places, devitrified tuffs. Included in this unit are the nonwelded tuffs at the base of the Tiva Canyon Tuff, the Yucca Mountain Tuff, the Pah Canyon Tuff, the nonwelded tuffs at the top of the Topopah Spring Tuff, and the associated bedded tuffs.

The TSw thermal/mechanical unit underlies the PTn. This unit is subdivided into three subunits based on volume of lithophysal cavities and the identification of the crystal-rich vitrophyre. The top subunit, TSw1, is lithophysae-rich and includes the top vitrophyre (crystal-rich member, vitric zone, densely welded subzone), the nonlithophysal zone, and the upper lithophysal zone. This upper subunit ranges from ~49 m to 113 m thick. The middle subunit, TSw2, is lithophysae-poor and consists of the middle nonlithophysal, lower lithophysal, and lower nonlithophysal zones. The TSw2 subunit ranges in thickness from 175 m to 229 m. The vitrophyre subunit (TSw3) at the base of TSw is ~7 m to 25 m thick.

Underlying the TSw unit is the CHn. This unit consists of the lower, nonwelded to partially welded portion of the Topopah Spring Tuff, the Calico Hills Formation, the underlying pre–Calico Hills bedded tuff, and the upper, nonwelded portions of the underlying Prow Pass Tuff of the Crater Flat Group.

As shown in Table 3-1, the TSw2 thermal/mechanical unit and the lower portion of the TSw1 thermal/mechanical unit have been identified as the proposed repository host horizon (CRWMS M&O, 1997d). This includes the lower portion of the Tptpul (upper lithophysal zone) lithostratigraphic unit, and the whole of the Tptpmn (middle nonlithophysal zone), Tptpll (lower lithophysal zone), and Tptpln (lower nonlithophysal zone) lithostratigraphic units.

Geographic Distribution of Data

The primary focus of the data collection program was the lithostratigraphic units constituting the potential repository horizon: the lower portion of the Tptpul (upper lithophysal zone), the Tptpmn (middle nonlithophysal zone), the Tptpll (lower lithophysal zone), and the Tptpln (lower nonlithophysal zone) (CRWMS M&O, 1997d).

Use of Data for Three-Dimensional Rock Properties Modeling

The Three-Dimensional (3-D) Integrated Site Model of Yucca Mountain, Nevada, consists of a geometric representation of selected rock units and structures, a geologic framework model, plus a set of rock properties and mineralogy models and data sets (CRWMS M&O, 2000). The framework model encompasses a 166 km^2 (64 mi^2) rectangle around Yucca Mountain and includes 50 rock units selected to meet the requirements of hydrologic and transport models and to represent the geology of Yucca Mountain. Modeled rock properties included matrix porosity, bulk "lithophysal" porosity, saturated hydraulic conductivity, density, and thermal conductivity.

Rautman (1995) and Rautman and McKenna (1997) implemented a two-dimensional (2-D) geostatistical model of porosity at Yucca Mountain and developed a correlation between that porosity model and thermal conductivity. It was assumed for modeling purposes that the spatial continuity of thermal conductivity was identical to those patterns exhibited by lithophysal porosity. This assumption was necessary because there are not enough actual measurements of thermal conductivity from which to develop a spatial continuity model on its own.

The result is a suite of thermal conductivity models that exhibit approximately the same cross-variable correlation with porosity as do the few measured data. It is anticipated that such a geostatistical modeling of thermal conductivity based on laboratory data from intact rock may be used in numerical programs to effectively simulate the rock mass thermal conductivity behavior. There may be some uncertainties in scaling thermal conductivity data from small intact rock sample testing to the repository or rock mass scale.

ROCK STRUCTURE

Rock Structural Data from Surface Mapping, Underground Mapping, and Geophysical Studies

The nature of jointed rock masses requires that both the frequency and characteristics of jointing are assessed along with in situ stress and intact rock strength. Jointing in the rock mass is often the primary factor controlling excavation stability. Joint and fracture data are incorporated into rock mass classification systems to estimate tunnel support requirements. Jointing can also have a significant influence on penetration rates of mechanical rock excavation equipment. Frequency and orientation of prominent joint sets are considered when estimating tunnel boring machine penetration rates.

Rock mass discontinuities are subdivided into groups according to their origin, because the origin of jointing in the project area influences the characteristics of the joint, such as roughness and planarity.

Joints within the Paintbrush Group have been subdivided into early cooling joints, later tectonic joints, and joints generated by erosional unloading. Each type of joint exhibits different characteristics that may impact trace length, connectivity, and orientation. Cooling joints and tectonic joints are similar in orientation but differ in surface roughness. Joints caused by erosional unloading have a different orientation and tend to be cross joints terminating at preexisting joints (Sweetkind and Williams-Stroud, 1996).

Cooling joints are identified in every unit that is at least moderately welded. They occur as two nearly orthogonal sets of steeply dipping joints, with a third, subhorizontal, irregular joint set. Cooling joints are characterized as having low surface roughness, with a joint roughness coefficient less than five. Tectonic joints have rougher surfaces (joint roughness coefficient greater than five) and occur throughout all units of the Paintbrush Group. The predominant strike is also N-S, with less frequent strike trends of NW and NE (CRWMS M&O, 1997d).

Joints caused by erosional unloading were variably oriented but trended predominantly E-W (Sweetkind and Williams-Stroud, 1996). Joint orientations appear to be related primarily to lithology, with a lesser association to major structures. The dominant dip of joints in all lithostratigraphic units is near vertical. This trend is moderated in the undifferentiated overburden (Tmr, Tpki) and the crystal-rich members of the Tiva Canyon Tuff and

Topopah Springs Tuff, where fractures with moderate dips were observed (CRWMS M&O, 1997d).

Multiple joint sets are well developed in the Tiva Canyon Tuff (CRWMS M&O, 1997d). All lithostratigraphic units in the Tiva Canyon Tuff exhibit the subhorizontal joint set to some degree, as well as two sets of near-vertical joints. Nonwelded tuff of the undifferentiated overburden and PTn thermal/mechanical units has fewer joints and generally only a single N-S–trending set (CRWMS M&O, 1997d).

The subhorizontal joint set occurs in the crystal-rich member of the Topopah Spring Tuff, which corresponds to the upper portion of the TSw1 thermal/mechanical unit. However, it was not observed in the upper lithophysal zone (Tptpul) of the TSw1 (Sweetkind and Williams-Stroud, 1996). Estimated rock quality in Tptpul is distinctly better than the equivalent unit of the Tiva Canyon Tuff (Tpcpul) due to much lower frequency of recorded joints as well as fewer joint sets (Sweetkind and Williams-Stroud, 1996).

Multiple joint sets occurred in the TSw2 thermal/mechanical unit, with dominant near-vertical joints striking N-S and NW. The subhorizontal joint set also occurs (Sweetkind and Williams-Stroud, 1996).

Very general information on fracture intensity and connectivity indicates that the highest joint frequencies and connectivities occur in the nonlithophysal units of the Tiva Canyon and Topopah Spring Tuffs (Sweetkind and Williams-Stroud, 1996). Nonwelded tuffs of the PTn have the lowest joint frequencies and lowest observed connectivities.

Geophysical studies have augmented bedrock geology structural studies and have contributed to the understanding of the spatial distribution of the geoengineering properties of the rocks at Yucca Mountain. Surface and borehole geophysical studies conducted over the period from 1979 to 1996 at Yucca Mountain are synthesized and summarized in Majer et al. (1996) and CRWMS M&O (1996b).

In addressing the spatial distribution of porosity (fracture porosity and matrix porosity) throughout the proposed repository, the geophysical model indicates that welding, lithophysal character and content, porosity, and mineralogy are variable over tens of meters. The geophysical model is not able to discriminate variation in porosity caused by fracturing from large-scale changes in matrix properties. However, in cases where there is good surface evidence for faulting, it appears that faulting and fracturing are the main causes for the variability in the geophysical data. The general results imply that the subsurface rock is very heterogeneous and complex at a rather small scale (Majer et al., 1996b).

Rock Structural Data from Boreholes

Structural data developed from cores include core recovery, locations of fractures and vugs, fracture characteristics, hardness, weathering, rock quality designation (RQD), and lithophysal and other voids.

Core Recovery

Core recovery can be used as a general indicator of relative rock quality. Substantial amounts of core in all lithostratigraphic units were either lost or recovered as rubble. Combining data from each of the boreholes, the amount of lost core for the Tptpmn lithostratigraphic unit was 15% of the total core length (Table 3-2). Rubble zones accounted for 20% of the total length. Total lost core and rubble ranged from 28.4% in the TCw to 49.2% in the TSw2 (Brechtel et al., 1995). In the lithophysal-rich portions of TCw and TSw1, the proportion of lost core and rubble zones was 32.6% and 51.2%, respectively.

To limit the introduction of water and chemical additives that are typically used in core drilling, a modified method with dry compressed air as the circulating medium was used. The effect of this dry drilling technique on the proportion of lost core and rubble was assessed by Brechtel et al. (1995), who concluded that the amount of lost core and rubble was related to the fractured nature of the welded tuffs and the presence of lithophysal voids, and not to the drilling technique.

The high proportion of lost core and rubble is attributed to the degree of small-scale fracturing of the welded rocks and the presence of other heterogeneities, such as lithophysae and vapor-phase alteration. Fracturing also may have been induced along a subhorizontal fabric or foliation by drilling. These heterogeneities had a large influence on core recovery and core quality, but had much less significance at the tunnel scale. Lost core and rubble zones did, however, limit description of core in places where individual features could not be reconstructed.

Rock Quality Designation (RQD)

While core recovery is related to the quality of rock encountered in a boring, it also is influenced to some degree by the drilling technique and type and size of core barrel used. The rock quality designation (RQD) (Deere, 1963) is a recovery ratio that provides an alternative estimate of in situ rock quality. This ratio is determined by considering only pieces of core that are at least 100 mm (4 in) long. The percentage ratio between the total length of such core recovered and the length of core

drilled on a given run is the rock quality designation, as follows (Deere, 1963):

$$RQD(\%) = \frac{\sum \text{Piece lengths} \geq 100 \text{ mm}}{\text{Interval length}} \qquad (3\text{-}1)$$

This index has been widely used as a general indicator of rock mass quality and is used as an input for determination of rock mass quality, such as rock mass rating (RMR) and rock mass quality index (Q), discussed later in this report. RMR and Q are indices that consider characteristics of the rock mass, such as the degree of jointing, strength, and groundwater condition, to classify the rock mass according to rock quality.

The RQD used for geotechnical design purposes considered all breaks in the core, including those identified by geological/geotechnical staff as drilling-induced and those indeterminate as to their natural or drilling-induced origin. Differentiation among types of breaks was difficult because of the general absence of mineralization along fractures and the existence of a subhorizontal fabric similar to foliation in some of the tuffs. The resulting rock quality designation, used for subsequent estimations of rock mass quality, is thus considered conservative because it indicates a lower rock quality than if only natural fractures had been considered. Attempts were made to filter the effects of drilling-induced fracturing. A parameter called enhanced–rock quality designation (RQD) was calculated (Brechtel et al., 1995), in which the fractures judged to be drilling-induced were not considered in RQD calculations, thus resulting in greater piece lengths. Enhanced-RQD values were 1.5–2.2 times higher than RQDs that considered all fractures. However, because of the uncertainty in fracture classification and because of the large proportion of lost core and rubble, the more conservative RQD was used for rock mass quality estimations.

RQD was generally not high in any unit because of the relatively low recovery of intact core, the high frequency of core fractures, and the consideration of drilling-induced mechanical breaks as fractures. Using the relative rock quality descriptions based on rock quality designation developed by Deere (1963), rock quality of core in the Tptpmn stratigraphic zone of the

Table 3-2. Summary of Core Recovery Data for Tptpmn

Boreholes	Total Logged (m)	Whole Core Recovered (m)	% of Total	Lost Core (m)	% of Total	Rubble Zones (m)	% of Total	Lost Core & Rubble (m)	% of Total
SD-7	36.6	21.2	57.9	3.9	10.7	11.5	31.4	15.4	42.1
SD-9	33.5	23.4	69.9	4.8	14.3	5.3	15.8	10.1	30.1
SD-12	39.6	27.4	69.2	3.5	8.8	8.7	22.0	12.2	30.8
NRG-6	30.5	21.8	71.5	2.7	8.9	6.0	19.7	8.7	28.5
NRG-7/7A	36.6	17.3	47.2	12.5	34.2	6.8	18.6	19.3	52.8
UZ-14	33.5	26.3	78.5	3.2	9.5	4.0	11.9	7.2	21.5
Total	210.3	137.4	65.3	30.6	14.5	42.3	20.1	72.9	34.7

TSw2 thermal/mechanical unit ranges from poor to very poor among the boreholes evaluated. Within the Tptpmn are two intervals of generally higher rock quality designation bounding a lower rock quality designation interval defined by the lithophysal subzone of the Tptpmn lithostratigraphic unit. The portion of the TSw2 unit cut by the Exploratory Studies Facility Main Drift is characterized by very low rock quality designations, and these low values are consistent from hole to hole. Based on rock quality designation values, core from the TCw and PTn thermal/mechanical units are classified as poor-quality rock, and core from the TSw1 is classified as very poor-quality rock (Brechtel et al., 1995). Locally, however, the nonlithophysal portion of the TSw1 is of higher quality and is classified as poor instead of very poor rock.

Based on the minimal tunnel support that was required in the Exploratory Studies Facility, rock quality is very much higher from an engineering standpoint than indicated by the low borehole RQD. This is related to the inclusion of mechanical and drilling-induced fractures along with natural fractures for RQD calculation, the apparent natural proclivity of the lithophysal zones to produce rubble zones at the core scale that do no significantly affect stability at the tunnel scale, and the use of dry drilling in fairly brittle rocks, which may have produced more fracturing than mud drilling. Because the recovered core was highly fractured, it was difficult to separate features that represented continuous joints from smaller-scale incipient cracks and heterogeneities in the rock (CRWMS M&O, 1997d).

Rock Weathering

Qualitative rock weathering descriptors were applied to describe the average condition of the core in each core run interval. These standard descriptors are based on recommendations of the International Society for Rock Mechanics (1981) (Table 3-3).

All rock from the Tptpmn in the TSw2 unit was either fresh or slightly weathered, constituting 51 and 49%, respectively, of the total recovered core (Kicker et al., 1996). A fairly high percentage of the core recovered from other welded tuff units, TCw and TSw1, was also fresh or slightly weathered, constituting ~60–70%. About 20% of the core length in the PTn unit was moderately weathered (Brechtel et al., 1995).

Rock Hardness

Rock hardness is a general descriptor of the strength of the rock material. Estimated rock hardness for recovered core was determined following the procedures in Sandia National Laboratories Technical Procedure TP-233, *Geotechnical Logging of Core from Examination of Core and Video Records*.

Rock in the Tptpmn of the TSw2 unit was generally classified from core logging as almost entirely hard (79%) to very hard (16%) (Kicker et al., 1996). Most of the rock in the other welded tuff units was estimated to be classified as very hard, hard, or moderately hard, with the highest percentage in the hard category (Table 3-4) (Brechtel et al., 1995). The nonwelded rock of the PTn unit was much softer than the welded tuffs, and approximately half of the core in this unit was estimated to be in the soft or very soft category (Brechtel et al., 1995).

Fractures

Fractures judged to have been induced by drilling generally were oriented normal to the core axis and were believed to have formed along subhorizontal planes of incipient weakness (Brechtel et al., 1995). These fractures accounted for 40–70% of the total fractures, depending on the thermal/mechanical unit. Only natural fractures and those indeterminate as to natural or drilling-induced origin were included in evaluations of fracture inclination, frequency, infillings, and roughness. Fractures judged to be drilling-induced were excluded (Brechtel et al., 1995).

Because of testing requirements that the borehole walls be kept dry and clear (i.e., protected from water and mudcake contamination), high-velocity, dry, compressed air was used as the drilling fluid instead of mud for core drilling. The effect of this drilling technique on fracturing in the core was difficult to establish because of limited data available for comparison (Brechtel et al., 1995). Brechtel et al. (1995) concluded that the amount of lost core and rubble was not caused by the dry drilling technique but was related to the fractured nature of the welded units and the presence of lithophysal voids. It is possible, however, that because of the brittle nature of these volcanic sequences, dry drilling may have caused increased fracturing relative to what might have occurred in mud drilling.

When fracture frequencies were corrected for sampling bias related to the relative orientation of the borehole with respect to fractures, the near-vertical fractures were most numerous. This agreed well with surface mapping observations (Brechtel et al., 1995).

Most fractures were clean, and fracture infillings, when observed, were rated as either "thin" or just "surface sheen." The most frequent infilling materials were identified as white crystalline, white noncrystalline, or black dendritic minerals.

Fracture surfaces in core samples were classified predominantly as having an irregular shape. The secondmost frequent fracture surface shapes were planar. Most fracture surfaces were rated as moderately rough.

Potential Key Blocks in Underground Excavations

Key blocks are rock wedges, formed by the intersection of geologic discontinuities and an excavation surface, that are kinematically able to move into the excavation. Key block analyses were performed to verify that ground support being installed in the Exploratory Studies Facility was adequate. Analyses were done prior to excavation of the Exploratory Studies Facility Main Drift to compare the size of potential key blocks that might occur in the North Ramp with the size of blocks projected for the Main Drift alignment, assuming that the joint sets encountered in the North Ramp also would be encountered in the Main Drift. In addition, analyses of a 100 m section of the North Ramp were performed to assess the importance of the subhorizontal vapor-phase parting structure that can be found in welded units.

Table 3-3. Rock Weathering Descriptions

Weathering Class	Log Abbreviation	Description
Fresh	F	Rock and fractures not oxidized or discolored; no separation of grains, change of texture, or solutioning
Slightly Weathered	S	Oxidized or discolored fractures and nearby rock; some dull feldspars; no separation of grains; minor leaching.
Moderately Weathered	M	Fractures and most of the rock oxidized or discolored; partial separation of grains; rusty or cloudy crystals; moderate leaching of soluble minerals.
Intensely Weathered	I	Fractures and rock totally oxidized or discolored; extensive clay alteration; leaching complete; extensive grain separation; rock is friable.
Decomposed	D	Grain separation and clay alteration complete.

Source: Sandia National Laboratories Technical Procedure SNL TP-233, *Geotechnical Logging of Core by Examination of Core and Video Records.*

Table 3-4. Estimated Rock Hardness Descriptions

Hardness Class	Log Abbreviation	Description
Extremely Hard	1	Cannot be scratched; chipped only with repeated heavy hammer blows.
Very Hard	2	Cannot be scratched; broken only with repeated hammer blows.
Hard	3	Scratched with heavy pressure; breaks with heavy hammer blow.
Moderately Hard	4	Scratched with light-moderate pressure; breaks with moderate hammer blow.
Moderately Soft	5	Grooved (1/16th inch) with moderate heavy pressure; breaks with light hammer blow.
Soft	6	Grooved easily with light pressure; scratched with fingernail; breaks with light-moderate manual pressure.
Very Soft	7	Readily gouged with fingernail; breaks with light pressure.
Soil-Like	8	Cohesive
Soil-Like	9	Noncohesive

Source: Sandia National Laboratories Technical Procedure SNL TP-233, *Geotechnical Logging of Core by Examination of Core and Video Records.*

The UNWEDGE (University of Toronto Rock Engineering Group, 1992) key block stability analysis program was used to rapidly estimate maximum potential key block sizes. The maximum potential key block size is the volume of the largest key block that could form, given an existing excavation configuration and a particular combination of joint set orientations. Specific individual joints cannot be positioned in specific locations. The program is nonprobabilistic, but determines the maximum potential block types and volumes based on the provided joint set data. It is limited to analysis of no more than three joint sets at a time. Input parameters include joint set orientation, joint spacing, cohesion and friction angle of each joint set, and rock unit weight.

The comparative analysis of the potential key blocks in the North Ramp and the Main Drift suggested that, given similar rock mass conditions, the Main Drift would have a higher frequency of key blocks because of the orientation of joints relative to the tunnel alignment.

A second analysis was performed to predict the maximum potential key block size that could form based on joint sets observed between stations 12+00 m and 13+00 m in the Exploratory Studies Facility. Although the subhorizontal joints in this interval could have had a negative impact on tunnel stability, most key blocks that were analytically predicted could be supported by the installed rock bolt support system. Two potential key blocks were found to have sufficient size to exceed the capacity of the rock bolt support system. However, no key blocks of this size were observed in the interval. The key block analysis must be considered conservative because of its inherent assumptions. Observations in the constructed Exploratory Studies Facility indicate that the installed supports are adequate, and no failures of significant rock wedges have been reported.

Additional key block analyses were performed as part of a study to analyze the deterioration of the rock mass surrounding the potential repository emplacement drifts (CRWMS M&O,

2000). A statistical description of the probable block sizes formed by fractures around the emplacement drifts was developed for each of the lithologic units of the repository host horizon. The change in drift profile resulting from progressive deterioration of the emplacement drifts was assessed both with and without backfill. Drift profiles were determined for four different time increments and three seismic levels (CRWMS M&O, 2000).

LABORATORY PROPERTIES OF INTACT ROCK

Physical Properties

Bulk and physical property measurements were performed on core specimens from NRG and SD boreholes. Additional measurements were performed on samples from the Single Heater Test block and the Drift Scale Test block of the Thermal Test Facility of the Exploratory Studies Facility. The data were collected under a qualified quality assurance program and include density, porosity, and mineralogy.

Density and Porosity

Density, a physical property defined as mass per unit volume at a specific temperature, can vary substantially within a rock mass because of variations in mineralogy, porosity, and welding. Average grain density is controlled by the composition of the rock, and variations in average grain density are attributable to variations in mineralogy and petrology. Porosity is a measure of the volume of voids in rock. It can be calculated from the relationship of average grain density and dry bulk density and also from the relationship of saturated bulk density and dry bulk density. Effective elastic constants, rock fracture, and rock rheological behavior are controlled in a large part by the size, shape, and distribution of pores.

Dry bulk density, saturated bulk density, and average grain density were measured for laboratory specimens of tuff prepared from borehole cores. Dry bulk density represents the mass measured in a dry condition divided by the volumetric measurement of the external dimensions of a test specimen. The Sandia National Laboratories procedure used, SNL TP-229, *Bulk Properties Determinations of Tuffaceous Rocks: Dry Bulk Density, Saturated Bulk Density, Average Grain Density, and Porosity,* generally adhered to the American Society for Testing and Materials Standards ASTM D854, *Standard Test Method for Specific Gravity of Soils* and ASTM C135, *Standard Test Method for True Specific Gravity of Materials by Water Immersion.* Test specimens recovered from the NRG boreholes were dried prior to testing, to a constant weight at 110 °C, and the dry bulk density was determined. Heating specimens at 110 °C is a deviation from the ASTM recommended procedures, because excessive temperatures may damage the specimen due to differential thermal expansion along grain boundaries. However, a series of thermal diagnostic tests, conducted as part of this study, showed that drying welded tuff specimens to 110 °C produced no noticeable damage. This was demonstrated by measuring the compressional and shear wave velocities before and after heating characteristic

welded tuff specimens. If microcracks had developed, the velocity would have decreased. The thermal cycling produced no reduction in velocity, indicating no thermally induced microcracking.

Once the dry bulk density was measured, each specimen was saturated with distilled water. A two-stage process was used for the saturation. The specimen was first pressure-saturated at 10 MPa for a minimum of 1 h, followed by a minimum of two vacuum-saturation cycles. Once the mass change for successive cycles had stabilized to within ±0.05%, the saturated bulk density was computed (Boyd et al., 1996a; CRWMS M&O, 1997d).

The average grain density was measured using the pycnometry method. This technique consists of a two-stage measurement. First, the mass of a dry, powdered specimen is measured. Next, the volume of the powder is determined. These two measurements are combined to compute the average grain density. Detailed procedures for the measurement are presented in Boyd et al. (1994). Total porosity, ϕ_T, was computed from the grain density, ρ_g, and the dry bulk density, ρ_{db}, using the expression (Boyd et al., 1994):

$$\phi_T = \frac{\rho_g - \rho_{db}}{\rho_g}. \tag{3-2}$$

Dry and saturated bulk densities, average grain density, and calculated porosity for specimens from the NRG boreholes are presented in CRWMS M&O (1997d), Martin et al. (1994, 1995) and Boyd et al. (1996a, 1996b). The data are summarized by thermal/mechanical unit in Tables 3-5, 3-6, 3-7, and 3-8. As shown, nonwelded rocks of the undifferentiated overburden and PTn units have significantly lower bulk density and higher porosity than rocks of welded tuff units TCw, TSw1, and TSw2. Because of the relatively uniform composition of the tuffs, average grain density shows only small variability among different rock units. Mean dry bulk density was 1.28 g/cm³ in both the nonwelded units' undifferentiated overburden, and PTn. Mean dry bulk density in the other, welded units was substantially higher and ranged from 2.12 g/cm³ to 2.35 g/cm³.

As part of the preheating, ambient characterization of hydrologic properties, density and porosity measurements were also made for core samples from wet-drilled and dry-drilled underground boreholes in the Single Heater Test block and the Drift Scale Test block of the Exploratory Studies Facility. Grab samples from the Observation Drift of the Drift Scale Test were also tested. All tested samples were from the TSw2 thermal/mechanical unit. Complete results and procedures are presented and discussed in CRWMS M&O (1996b) and CRWMS M&O (1997g). Results are generally consistent with TSw2 results from surface-based boreholes (Table 3-9). As shown, mean bulk density ranged from 2.20 to 2.26 g/cm³, mean particle density ranged from 2.49 to 2.51 g/cm³, and mean porosity ranged from 9.3 to 12.5%.

Mineralogy

The mineralogy and petrology of tuffs affect the thermal and mechanical geoengineering properties, and thus help characterize the rock mass at the site. Because large quantities of heat will be generated by the nuclear waste packages, it is essential that heat-transfer properties (such as thermal conductivity, thermal expansion, and thermal heat capacity) be understood so that repository performance can be assessed. Thermal properties are largely a function of mineralogy, and so an understanding of these properties requires that mineralogy be determined.

The abundance of minerals that could affect the thermal/mechanical behavior of the rock at elevated temperatures is of particular interest. These minerals include cristobalite, which undergoes a phase transition and volume change at elevated temperatures, and smectites and zeolites, which dehydrate at elevated temperatures with accompanying volume reduction. Other mineralogical effects associated with temperature changes include the dissolution and precipitation of silica and the dehydration of volcanic glass. All these effects in turn affect the thermal and mechanical properties of the rock mass. The purpose of mineralogical and petrological analyses is to evaluate correlations between the thermal/mechanical behavior of rock samples and their mineralogical components and texture.

In brief, both the Tiva Canyon Tuff and the Topopah Spring Tuff are petrologically zoned ignimbrites with crystal-poor rhyolitic units at the base and crystal-rich quartz latite units at the top. Both units have devitrified zones, welding zones, and secondary crystallization. A suite of 97 samples from borehole NRG-6 was studied to understand the mineralogy and petrology of samples tested for thermal expansion, thermal conductivity, and mechanical properties (CRWMS M&O, 1997d). Borehole NRG-6 was selected because it provides a relatively complete stratigraphic section from the lower part of the Tiva Canyon Tuff through most of the lower lithophysal zone of the Topopah Spring Tuff of the Paintbrush Group (Table 3-10). Depths ranged from 6.77 to 330.7 m (22.2–1085.0 ft).

The Single Heater Test and the Drift Scale Test in the Thermal Test Facility related mineralogy to thermal, mechanical, chemical, and hydrological processes. Analyses were performed for 10 samples from the Single Heater Test region of the Thermal Test Facility. The analyses used the matrix flushing method of Chung (1974) and an internal intensity standard (corundum) to quantify the phases present.

Mineral abundances were also determined for 17 samples from the Drift Scale Test block of the Thermal Test Facility of the Exploratory Studies Facility using the Rietveld method of whole X-ray pattern fitting (Snyder and Bish, 1989; Young, 1993). Complete procedures and results are described in Roberts and Viani (1997) and CRWMS M&O (1997g). The method appears to yield better precision than the matrix flushing method of Chung (1974) used for Single Heater Test samples (Table 3-11).

The most obvious physical divisions are between welded and nonwelded tuffs, which are reflected in the variation in thermal properties. Invariably in the samples from NRG-6, the welded intervals show significant compaction, resulting in the destruction of primary porosity. With few exceptions and above the water table, the nonwelded intervals retain their primary glassy or vitric character and significant porosity. These intervals rarely correspond to stratigraphic rock unit boundaries because the upper and lower layers of eruptive units (upon which the stratigraphic units are defined) are typically nonwelded while the interiors of the units are welded (CRWMS M&O, 1997d).

Minerals identified from X-ray powder diffraction on 23 devitrified samples from NRG-6 included the silica phases (cristobalite, quartz, tridymite), feldspar (plagioclase and Na- and K-rich alkali feldspar), and a clay mineral (primarily illite). Other minerals were noted in thin sections but were not identifiable by X-ray diffraction because of their low abundances. These include small amounts of quartz present as phenocrysts, biotite, amphibole, and Fe-Ti oxides (CRWMS M&O, 1997d).

Cristobalite is present in virtually all devitrified samples and is the dominant silica phase in most samples. Quartz is absent in all samples shallower than 139.8 m (458.7 ft) and only appears abundantly below 197.7 m (648.6 ft). Tridymite is intermittently present in samples at 129.6 m (425.3 ft) or less in depth and was not identified in any quartz-bearing samples. Clay (illite) is identified in many, but not all, samples in the same interval from which tridymite was identified. It also was noted that silica phase variations tend to cross the lithostratigraphic boundaries defined by Geslin et al. (1995) in their lithostratigraphic classification of rocks from borehole NRG-6.

For samples from the Single Heater Test, quartz and cristobalite average ~8 and 19%, respectively. Smectite is found in all the samples, but is generally <3%. The greatest variability is shown in the abundances of albite and sanidine. Clinoptilolite is present in small amounts in three of the samples. Because the method used to estimate mineral abundances is not normalized to 100%, the total mineral abundances average of the total is 96 ± 4%. This low value suggests that an unidentified phase may be present. The observed mineralogy is consistent with previous measurements of mineral abundances in core samples of Topopah Spring devitrified tuff (Bish and Chipera, 1986).

All samples from the Drift Scale Test block were from the TSw2 thermal/mechanical unit (Table 3-12) (CRWMS M&O, 1997g). The total abundance of the silica polymorphs is fairly uniform, although the cristobalite component varies from 4 to 31%, suggesting potential variability in thermal/mechanical properties at the temperatures at which cristobalite undergoes a phase change. In most samples, albite, sanidine, and cristobalite are the dominant phases, with lesser amounts of quartz. Tridymite is significant in three samples, with cristobalite being less abundant in these samples. Zeolite phases were observed in three samples, clinoptilolite in two samples, and stilbite in one sample. Compared to the analyses of samples from the Single Heater Test block (Roberts and Viani, 1997; Viani and Roberts, 1996), these samples show similar total silica polymorph content (between 33 and 41%). The quartz and cristobalite contents appear to be inversely related.

Mineral hardness or abrasivity is commonly used to predict rates of Tunnel Boring Machine cutter wear. Measurements of rock abrasivity, such as the Cercher Abrasivity test, have not been performed on Yucca Mountain rock samples, but abrasivity can be

Table 3-5. Summary of Dry Bulk Density (g/cm³), Thermal/Mechanical Units

Core Hole	UO			TCw			TCw-N			TCw-L		
	Mean	Std. Dev.	Count	Mean	Std. Dev.	Count	Mean	Std. Dev.	Count	Mean	Std. Dev.	Count
NRG-2	N/A	N/A	0	2.34	0.01	17	2.34	0.01	17	N/A	N/A	0
NRG-2A	1.23	0.06	12	1.86	0.22	19	1.86	0.22	19	N/A	N/A	0
NRG-2B	1.33	0.11	10	N/A	N/A	0	N/A	N/A	0	N/A	N/A	0
NRG-3	N/A	N/A	0	2.13	0.19	45	2.13	0.21	32	2.14	0.10	13
NRG-4	N/A	N/A	0	N/A	N/A	0	N/A	N/A	0	N/A	N/A	0
NRG-5	N/A	N/A	0	N/A	N/A	0	N/A	N/A	0	N/A	N/A	0
NRG-6	N/A	N/A	0	2.28	0.15	29	2.17	0.19	11	2.35	0.03	18
NRG-7	N/A	N/A	0	2.24	0.10	11	2.24	0.10	11	N/A	N/A	0
SD-9	N/A	N/A	0	N/A	N/A	0	N/A	N/A	0	N/A	N/A	0
SD-12	N/A	N/A	0	N/A	N/A	0	N/A	N/A	0	N/A	N/A	0
ALL HOLES												
TOTAL	1.28	0.10	22	2.16	0.22	121	2.13	0.24	90	2.26	0.12	31
Range	1.13 to 1.53			1.45 to 2.37			1.45 to 2.36			1.95 to 2.37		
Mean	1.28			2.16			2.13			2.26		
Std. Dev.	0.10			0.22			0.24			0.12		
N	22			121			90			31		

Table 3-5. Summary of Dry Bulk Density (g/cm³), Thermal/Mechanical Units (Continued)

Core Hole	PTn			TSw1			TSw1-N			TSw1-L		
	Mean	Std. Dev.	Count	Mean	Std. Dev.	Count	Mean	Std. Dev.	Count	Mean	Std. Dev.	Count
NRG-2	N/A	N/A	0	N/A	N/A	0	N/A	N/A	0	N/A	N/A	0
NRG-2A	N/A	N/A	0	N/A	N/A	0	N/A	N/A	0	N/A	N/A	0
NRG-2B	N/A	N/A	0	N/A	N/A	0	N/A	N/A	0	N/A	N/A	0
NRG-3	N/A	N/A	0	N/A	N/A	0	N/A	N/A	0	N/A	N/A	0
NRG-4	1.26	0.17	15	2.13	0.08	51	2.14	0.07	46	2.03	0.04	5
NRG-5	N/A	N/A	0	2.04	N/A	1	N/A	N/A	0	2.04	N/A	1
NRG-6	1.22	0.20	14	2.22	0.04	43	2.22	0.04	37	2.20	0.03	6
NRG-7	1.33	0.24	28	2.15	0.08	81	2.18	0.06	52	2.12	0.09	29
SD-9	N/A	N/A	0	N/A	N/A	0	N/A	N/A	0	N/A	N/A	0
SD-12	N/A	N/A	0	N/A	N/A	0	N/A	N/A	0	N/A	N/A	0
ALL HOLES												
TOTAL	1.28	0.21	57	2.16	0.08	176	2.18	0.07	135	2.12	0.09	41
Range	1.00 to 1.78			1.94 to 2.40			1.98 to 2.40			1.94 to 2.26		
Mean	1.28			2.16			2.18			2.12		
Std. Dev.	0.21			0.08			0.07			0.09		
N	57			176			135			41		

Table 3-5. Summary of Dry Bulk Density (g/cm³), Thermal/Mechanical Units (Continued)

Core Hole	TSw2 Mean	TSw2 Std. Dev.	TSw2 Count	TSw3 Mean	TSw3 Std. Dev.	TSw3 Count	CHn1 Mean	CHn1 Std. Dev.	CHn1 Count
NRG-2	N/A	N/A	0	N/A	N/A	0	N/A	N/A	0
NRG-2A	N/A	N/A	0	N/A	N/A	0	N/A	N/A	0
NRG-2B	N/A	N/A	0	N/A	N/A	0	N/A	N/A	0
NRG-3	N/A	N/A	0	N/A	N/A	0	N/A	N/A	0
NRG-4	N/A	N/A	0	N/A	N/A	0	N/A	N/A	0
NRG-5	2.26	0.04	13	N/A	N/A	0	N/A	N/A	0
NRG-6	2.25	0.06	27	N/A	N/A	0	N/A	N/A	0
NRG-7	2.27	0.09	72	2.35	0.00	4	N/A	N/A	0
SD-9	N/A	N/A	0	N/A	N/A	0	N/A	N/A	0
SD-12	N/A	N/A	0	N/A	N/A	0	N/A	N/A	0
ALL HOLES									
TOTAL	2.27	0.08	112	2.35	0.00	4			
Range	1.84 to 2.42			2.34 to 2.35					
Mean	2.27			2.35					
Std. Dev.	0.08			0.00					
N	112			4					

N—Number of samples; N/A—not applicable; Std. Dev—sample standard deviation

NOTE: The TCw and TSw1 thermal/mechanical units have been subdivided as follows:

TCw-N represents nonlithophysal zones of the TCw unit.
TCw-L represents lithophysal zones of the TCw unit.
TSw1-N represents nonlithophysal zones of the TSw1 unit.
TSw1-L represents lithophysal zones of the TSw1 unit.

Table 3-6. Saturated Bulk Density (g/cm³) by Thermal/Mechanical Unit

Core Hole	UO (Tuff "X")			TCw			TCw-N			TCw-L		
	Mean	Std. Dev.	Count	Mean	Std. Dev.	Count	Mean	Std. Dev.	Count	Mean	Std. Dev.	Count
NRG-2	N/A	N/A	0	2.40	0.01	17	2.40	0.01	17	N/A	N/A	0
NRG-2A	1.66	0.05	13	2.11	0.12	19	2.11	0.12	19	N/A	N/A	0
NRG-2B	1.73	0.08	8	N/A	N/A	0	N/A	N/A	0	N/A	N/A	0
NRG-3	N/A	N/A	0	2.27	0.12	45	2.27	0.13	32	2.26	0.08	13
NRG-4	N/A	N/A	0	N/A	N/A	0	N/A	N/A	0	N/A	N/A	0
NRG-5	N/A	N/A	0	N/A	N/A	0	N/A	N/A	0	N/A	N/A	0
NRG-6	N/A	N/A	0	2.36	0.12	26	2.28	0.16	10	2.41	0.01	16
NRG-7	N/A	N/A	0	2.34	0.06	11	2.34	0.06	11	N/A	N/A	0
SD-9	N/A	N/A	0	N/A	N/A	0	N/A	N/A	0	N/A	N/A	0
SD-12	N/A	N/A	0	2.39	0.03	13	2.39	0.03	13	N/A	N/A	0
ALL HOLES												
TOTAL	1.69	0.07	21	2.30	0.14	131	2.29	0.14	102	2.34	0.09	29
Range	1.56 to 1.85			1.87 to 2.43			1.87 to 2.43			2.10 to 2.42		
Mean	1.69			2.30			2.29			2.34		
Std. Dev.	0.07			0.14			0.14			0.09		
N	21			131			102			29		

Table 3-6. Saturated Bulk Density (g/cm³) by Thermal/Mechanical Unit (Continued)

Core Hole	PTn			TSw1			TSw1-N			TSw1-L		
	Mean	Std. Dev.	Count	Mean	Std. Dev.	Count	Mean	Std. Dev.	Count	Mean	Std. Dev.	Count
NRG-2	N/A	N/A	0	N/A	N/A	0	N/A	N/A	0	N/A	N/A	0
NRG-2A	N/A	N/A	0	N/A	N/A	0	N/A	N/A	0	N/A	N/A	0
NRG-2B	N/A	N/A	0	N/A	N/A	0	N/A	N/A	0	N/A	N/A	0
NRG-3	N/A	N/A	0	N/A	N/A	0	N/A	N/A	0	N/A	N/A	0
NRG-4	1.71	0.12	15	2.28	0.05	51	2.28	0.05	46	2.21	0.03	5
NRG-5	N/A	N/A	0	2.23	N/A	1	N/A	N/A	0	2.23	N/A	1
NRG-6	1.82	0.00	2	2.33	0.03	40	2.34	0.03	34	2.32	0.02	6
NRG-7	1.75	0.15	28	2.29	0.06	72	2.30	0.04	46	2.26	0.07	26
SD-9	1.71	0.10	6	2.32	0.01	4	2.32	0.01	3	2.31	N/A	1
SD-12	1.78	0.17	5	2.30	0.03	8	2.30	0.03	8	N/A	N/A	0
ALL HOLES												
TOTAL	1.74	0.13	56	2.30	0.05	176	2.31	0.05	137	2.26	0.07	39
Range	1.56 to 2.04			2.12 to 2.46			2.16 to 2.46			2.12 to 2.37		
Mean	1.74			2.30			2.31			2.26		
Std. Dev.	0.13			0.05			0.05			0.07		
N	56			176			137			39		

Table 3-6. Saturated Bulk Density (g/cm^3) by Thermal/Mechanical Unit (Continued)

Core Hole	TSw2 Mean	TSw2 Std. Dev.	TSw2 Count	TSw3 Mean	TSw3 Std. Dev.	TSw3 Count	CHn1 Mean	CHn1 Std. Dev.	CHn1 Count
NRG-2	N/A	N/A	0	N/A	N/A	0	N/A	N/A	0
NRG-2A	N/A	N/A	0	N/A	N/A	0	N/A	N/A	0
NRG-2B	N/A	N/A	0	N/A	N/A	0	N/A	N/A	0
NRG-3	N/A	N/A	0	N/A	N/A	0	N/A	N/A	0
NRG-4	N/A	N/A	0	N/A	N/A	0	N/A	N/A	0
NRG-5	2.36	0.03	13	N/A	N/A	0	N/A	N/A	0
NRG-6	2.36	0.04	27	N/A	N/A	0	N/A	N/A	0
NRG-7	2.38	0.03	55	2.35	0.01	3	N/A	N/A	0
SD-9	2.37	0.03	28	N/A	N/A	0	1.87	0.03	3
SD-12	2.39	0.04	10	2.36	0.00	3	N/A	N/A	0
ALL HOLES									
TOTAL	2.37	0.03	133	2.36	0.01	6	1.87	0.03	3
Range	2.26 to 2.45			2.35 to 2.37			1.85 to 1.90		
Mean	2.37			2.36			1.87		
Std. Dev.	0.03			0.01			0.03		
N	133			6			3		

N—Number of samples; N/A—not applicable; Std. Dev—sample standard deviation

NOTE: The TCw and TSw1 thermal/mechanical units have been subdivided as follows:
TCw-N represents nonlithophysal zones of the TCw unit.
TCw-L represents lithophysal zones of the TCw unit.
TSw1-N represents nonlithophysal zones of the TSw1 unit.
TSw1-L represents lithophysal zones of the TSw1 unit.

Table 3-7. Summary of Average Grain Density (g/cm³), Thermal/Mechanical Units

Core Hole	UO Mean	UO Std. Dev.	UO Count	TCw Mean	TCw Std. Dev.	TCw Count	TCw-N Mean	TCw-N Std. Dev.	TCw-N Count	TCw-L Mean	TCw-L Std. Dev.	TCw-L Count
NRG-2	2.44	0.01	3	2.53	0.01	17	2.53	0.01	17	N/A	N/A	0
NRG-2A	2.34	0.03	12	2.55	0.04	15	2.55	0.04	15	N/A	N/A	0
NRG-2B	2.34	0.02	10	N/A	N/A	0	N/A	N/A	0	N/A	N/A	0
NRG-3	N/A	N/A	0	2.53	0.03	26	2.54	0.03	16	2.52	0.01	10
NRG-4	N/A	N/A	0	N/A	N/A	0	N/A	N/A	0	N/A	N/A	0
NRG-5	N/A	N/A	0	N/A	N/A	0	N/A	N/A	0	N/A	N/A	0
NRG-6	N/A	N/A	0	2.49	0.02	11	2.48	0.03	6	2.50	0.01	5
NRG-7	N/A	N/A	0	2.50	0.01	7	2.50	0.01	7	N/A	N/A	0
SD-9	N/A	N/A	0	N/A	N/A	0	N/A	N/A	0	N/A	N/A	0
SD-12	N/A	N/A	0	N/A	N/A	0	N/A	N/A	0	N/A	N/A	0
ALL HOLES												
TOTAL	2.35	0.04	25	2.53	0.03	76	2.53	0.04	61	2.51	0.02	15
Range	2.30 to 2.46			2.44 to 2.61			2.44 to 2.61			2.49 to 2.55		
Mean	2.35			2.53			2.53			2.51		
Std. Dev.	0.04			0.03			0.04			0.02		
N	25			76			61			15		

Table 3-7. Summary of Average Grain Density (g/cm³), Thermal/Mechanical Units (Continued)

Core Hole	PTn Mean	PTn Std. Dev.	PTn Count	TSw1 Mean	TSw1 Std. Dev.	TSw1 Count	TSw1-N Mean	TSw1-N Std. Dev.	TSw1-N Count	TSw1-L Mean	TSw1-L Std. Dev.	TSw1-L Count
NRG-2	N/A	N/A	0	N/A	N/A	0	N/A	N/A	0	N/A	N/A	0
NRG-2A	N/A	N/A	0	N/A	N/A	0	N/A	N/A	0	N/A	N/A	0
NRG-2B	N/A	N/A	0	N/A	N/A	0	N/A	N/A	0	N/A	N/A	0
NRG-3	N/A	N/A	0	N/A	N/A	0	N/A	N/A	0	N/A	N/A	0
NRG-4	2.41	0.09	14	2.56	0.02	26	2.57	0.02	22	2.54	0.00	4
NRG-5	N/A	N/A	0	2.53	0.01	3	N/A	N/A	0	2.53	0.01	3
NRG-6	2.42	0.04	12	2.55	0.03	48	2.56	0.02	30	2.52	0.02	17
NRG-7	2.35	0.06	28	2.55	0.02	47	2.56	0.02	29	2.54	0.01	18
SD-9	N/A	N/A	0	N/A	N/A	0	N/A	N/A	0	N/A	N/A	0
SD-12	N/A	N/A	0	N/A	N/A	0	N/A	N/A	0	N/A	N/A	0
ALL HOLES												
TOTAL	2.38	0.08	54	2.55	0.02	124	2.56	0.02	81	2.53	0.01	42
Range	2.24 to 2.65			2.50 to 2.60			2.50 to 2.60			2.50 to 2.56		
Mean	2.38			2.55			2.56			2.53		
Std. Dev.	0.08			0.02			0.02			0.01		
N	54			124			81			42		

Table 3-7. Summary of Average Grain Density (g/cm^3), Thermal/Mechanical Units (Continued)

Core Hole	TSw2			TSw3			CHn1		
	Mean	Std. Dev.	Count	Mean	Std. Dev.	Count	Mean	Std. Dev.	Count
NRG-2	N/A	N/A	0	N/A	N/A	0	N/A	N/A	0
NRG-2A	N/A	N/A	0	N/A	N/A	0	N/A	N/A	0
NRG-2B	N/A	N/A	0	N/A	N/A	0	N/A	N/A	0
NRG-3	N/A	N/A	0	N/A	N/A	0	N/A	N/A	0
NRG-4	N/A	N/A	0	N/A	N/A	0	N/A	N/A	0
NRG-5	2.55	0.02	24	N/A	N/A	0	N/A	N/A	0
NRG-6	2.55	0.03	28	N/A	N/A	0	N/A	N/A	0
NRG-7	2.56	0.03	57	2.37	0.01	3	N/A	N/A	0
SD-9	N/A	N/A	0	N/A	N/A	0	N/A	N/A	0
SD-12	N/A	N/A	0	N/A	N/A	0	N/A	N/A	0
ALL HOLES									
TOTAL	2.55	0.03	109	2.37	0.01	3			
Range	2.42 to 2.61			2.37 to 2.38					
Mean	2.55			2.37					
Std. Dev.	0.03			0.01					
N	109			3					

N—Number; N/A—not applicable; Std. Dev—sample standard deviation

NOTE: The TCw and TSw1 thermal/mechanical units have been subdivided as follows:
TCw-N represents nonlithophysal zones of the TCw unit.
TCw-L represents lithophysal zones of the TCw unit.
TSw1-N represents nonlithophysal zones of the TSw1 unit.
TSw1-L represents lithophysal zones of the TSw1 unit.

Table 3-8. Summary of Porosity (%), Thermal/Mechanical Units

Core Hole	UO			TCw			TCw-N			TCw-L		
	Mean	Std. Dev.	Count	Mean	Std. Dev.	Count	Mean	Std. Dev.	Count	Mean	Std. Dev.	Count
NRG-2	N/A	N/A	0	7.45	0.66	17	7.45	0.66	17	N/A	N/A	0
NRG-2A	46.81	2.07	15	27.48	9.16	19	27.48	9.16	19	N/A	N/A	0
NRG-2B	42.90	4.33	10	N/A	N/A	0	N/A	N/A	0	N/A	N/A	0
NRG-3	N/A	N/A	0	15.71	8.16	44	16.01	9.42	31	14.98	3.97	13
NRG-4	N/A	N/A	0	N/A	N/A	0	N/A	N/A	0	N/A	N/A	0
NRG-5	N/A	N/A	0	N/A	N/A	0	N/A	N/A	0	N/A	N/A	0
NRG-6	N/A	N/A	0	8.56	5.57	29	12.24	7.12	12	5.96	1.40	17
NRG-7	N/A	N/A	0	11.13	5.08	12	11.13	5.08	12	N/A	N/A	0
SD-9	N/A	N/A	0	N/A	N/A	0	N/A	N/A	0	N/A	N/A	0
SD-12	N/A	N/A	0	N/A	N/A	0	N/A	N/A	0	N/A	N/A	0
ALL HOLES												
TOTAL	45.24	3.66	25	14.23	9.48	121	15.67	10.12	91	9.87	5.32	30
Range	35.20 to 51.00			4.80 to 44.50			5.20 to 44.50			4.80 to 22.60		
Mean	45.24			14.23			15.67			9.87		
Std. Dev.	3.66			9.48			10.12			5.32		
N	25			121			91			30		

Table 3-8. Summary of Porosity (%), Thermal/Mechanical Units (Continued)

Core Hole	PTn			TSw1			TSw1-N			TSw1-L		
	Mean	Std. Dev.	Count	Mean	Std. Dev.	Count	Mean	Std. Dev.	Count	Mean	Std. Dev.	Count
NRG-2	N/A	N/A	0	N/A	N/A	0	N/A	N/A	0	N/A	N/A	0
NRG-2A	N/A	N/A	0	N/A	N/A	0	N/A	N/A	0	N/A	N/A	0
NRG-2B	N/A	N/A	0	N/A	N/A	0	N/A	N/A	0	N/A	N/A	0
NRG-3	N/A	N/A	0	N/A	N/A	0	N/A	N/A	0	N/A	N/A	0
NRG-4	47.97	5.98	15	16.71	2.87	51	16.36	2.75	46	19.86	1.97	5
NRG-5	N/A	N/A	0	19.20	N/A	1	N/A	N/A	0	19.20	N/A	1
NRG-6	49.65	7.45	16	15.74	5.36	56	13.28	1.96	37	20.67	6.77	18
NRG-7	44.65	9.33	33	15.52	3.61	95	14.36	2.60	54	17.05	4.17	41
SD-9	N/A	N/A	0	N/A	N/A	0	N/A	N/A	0	N/A	N/A	0
SD-12	N/A	N/A	0	N/A	N/A	0	N/A	N/A	0	N/A	N/A	0
ALL HOLES												
TOTAL	46.68	8.38	64	15.90	4.03	203	14.74	2.77	137	18.30	5.11	65
Range	27.90 to 59.40			6.80 to 32.10			6.80 to 22.80			10.60 to 32.10		
Mean	46.68			15.90			14.74			18.30		
Std. Dev.	8.38			4.03			2.77			5.11		
N	64			203			137			65		

Table 3-8. Summary of Porosity (%),Thermal/Mechanical Units (Continued)

Core Hole	TSw2			TSw3			CHn1		
	Mean	Std. Dev.	Count	Mean	Std. Dev.	Count	Mean	Std. Dev.	Count
NRG-2	N/A	N/A	0	N/A	N/A	0	N/A	N/A	0
NRG-2A	N/A	N/A	0	N/A	N/A	0	N/A	N/A	0
NRG-2B	N/A	N/A	0	N/A	N/A	0	N/A	N/A	0
NRG-3	N/A	N/A	0	N/A	N/A	0	N/A	N/A	0
NRG-4	N/A	N/A	0	N/A	N/A	0	N/A	N/A	0
NRG-5	11.00	2.16	24	N/A	N/A	0	N/A	N/A	0
NRG-6	12.01	3.17	38	N/A	N/A	0	N/A	N/A	0
NRG-7	11.05	3.13	81	1.15	0.31	4	N/A	N/A	0
SD-9	N/A	N/A	0	N/A	N/A	0	N/A	N/A	0
SD-12	N/A	N/A	0	N/A	N/A	0	N/A	N/A	0
ALL HOLES									
TOTAL	11.30	3.01	143	1.15	0.31	4			
Range	3.80 to 27.70			0.70 to 1.40					
Mean	11.30			1.15					
Std. Dev.	3.01			0.31					
N	143			4					

N—Number; N/A—not applicable; Std. Dev—sample standard deviation

NOTE: The TCw and TSw1 thermal/mechanical units have been subdivided as follows:
TCw-N represents nonlithophysal zones of the TCw unit.
TCw-L represents lithophysal zones of the TCw unit.
TSw1-N represents nonlithophysal zones of the TSw1 unit.
TSw1-L represents lithophysal zones of the TSw1 unit.

Table 3-9.   Summary of Porosity and Density Measurements from the Thermal Test Facility of the Exploratory Studies Facility (TSw2 Thermal/Mechanical Unit)

Single Heater Test Borehole Summary (Wet-Drilled)			
	Porosity (%)	Bulk Density (g/cm^3)	Particle Density (g/cm^3)
Single Heater Test Average	12.53	2.20	2.51
Standard Deviation	3.89	0.09	0.017

Drift Scale Test PERM Borehole Summary (Dry-Drilled)			
	Porosity (%)	Bulk Density (g/cm^3)	Particle Density (g/cm^3)
Drift Scale Test PERM Average	10.62	2.25	2.51
Standard Deviation	1.14	0.02	0.01

DST Heated Drift Borehole Summary (Wet Excavation)			
	Porosity (%)	Bulk Density (g/cm^3)	Particle Density (g/cm^3)
Drift Scale Test Heated Drift Average	12.54	2.20	2.51
Standard Deviation	2.75	0.06	0.01

Observation Drift Grab Sample Summary (Wet-Excavation)			
	Porosity (%)	Bulk Density (g/cm^3)	Particle Density (g/cm^3)
Observation Drift Average	9.34	2.26	2.49
Standard Deviation	0.91	0.02	0.02

PERM = permeability borehole

Source: CRWMS M&O (1997g)

inferred from mineral composition. The Tunnel Boring Machine disc cutter steel has a hardness of ~6.5 on the Moh's scale of hardness. Rock with a large percentage of minerals of hardness equal to or exceeding 6.5 would produce faster cutter wear than rock with a smaller percentage of hard minerals. The percentage of quartz, with a Moh's hardness of 7, is typically used for cutter wear prediction. Generally, a quartz content less than 20% would indicate low cutter wear, between 20% and 50% would indicate significant cutter wear, and greater than 50% would indicate high cutter wear. The estimated combined percentage of quartz, cristobalite, and tridymite (which also have hardnesses of 7) is typically between ~1%5 and 30% for borehole NRG-6 and the Single Heater Test samples and between 33% and 40% for the Drift Scale Test samples (Tables 3-10, 3-11, and 3-12).

Thermal Properties

Thermal Conductivity

Thermal conductivity is a measure of the ability of a material to transmit heat, and so relates to the ability of the host rock to conduct heat away from nuclear waste containers. Thus, thermal conductivity is an important parameter for numerically simulating the transient temperature field from heat generated by emplaced radioactive waste.

Characterization of the thermal conductivities of Yucca Mountain Tuffs has been ongoing since 1980 (Lappin, 1980). Data published through 1988 were reviewed by Nimick (1989), who summarized data from boreholes G-1, G-2, GU-3, and G-4. The reports summarized in Nimick (1989) include Lappin et al. (1982), Lappin and Nimick (1985), Nimick and Lappin (1985), and Nimick et al. (1988). Nimick (1990a, 1990b) presented further analyses of data for the welded, devitrified portion of the Topopah Spring Tuff (units TSw1 and TSw2) and experimental results for the units overlying and including CHn2. Additional measurements of thermal conductivity were performed on core specimens from Yucca Mountain by Sass et al. (1988) using a needle-probe technique. However, the uncertainty in moisture contents for these specimens is discussed in Nimick (1990a, 1990b).

Test specimens were right circular cylinders, ~12.7 mm in length and 50.8 mm in diameter. Moisture contents were either air dry (as received), oven dry, vacuum saturated, or partially saturated (intermediate between air dry and vacuum saturated). Tests were conducted using the guarded heat flow meter over the temperature range of 30–300 °C. The test specimen was placed between two heater plates controlled at different temperatures, and the heat flow was measured. Radial heat flow losses were minimized using a cylindrical guard heater. The test procedure is given in SNL TP-202, *Measurement of Thermal Conductivity*

Table 3-10. **Mineral Abundances for 10 Samples from the Single Heater Test Block of the Exploratory Studies Facility**

Sample	Quartz	Cristobalite	Albite	Sanidine	Smectite	Clinoptilolite	Mica	Total
BJ-1-1.0-B ymp014[1]	10	18	41	26	tr[1]	nd[3]	tr	95
BJ-1-10.0-B ymp019	8	19	38	25	3	nd	tr	93
BJ-1-18.0B ymp011a ymp011a[4]	8 8	18 17	40 49	26 25	3 2	nd nd	nd nd	95 101
MPBX-3-12.5-B ymp018 ymp018a	11 12	18 18	33 34	32 28	3 3	4 6	<1 <1	101 101
MPBX-4-5.0-B ymp015	4	24	44	22	2	nd	nd	96
H1-0.6-C ymp013	8	17	43	25	tr	2	<1	95
H1-11.3-C ymp012	9	19	38	24	3	nd	tr	93
H1-11.6-C ymp016	7	19	37	25	3	nd	tr	91
H1-22.2-B ymp017 ymp017a	9 8	20 20	34 36	31 24	3 3	2 1	<1 tr	99 92

(1) The identification ymp### refers to computer filenames associated with the diffraction pattern data.
(2) tr - trace amount; the phase is present, but the peak intensities are too small to measure an integrated intensity.
(3) nd - not detected.
(4) Trailing "a" refers to duplicate sample.

of Geologic Samples by the Guarded-Heat-Flow Meter Method. The most applicable ASTM standard, one which relies on the same type of measurement equipment, is ASTM F433-77, *Standard Practice for Evaluating Thermal Conductivity of Gasket Materials.* There were no significant differences between the Sandia National Laboratories (SNL) and ASTM procedures.

Variation of thermal conductivity with saturation for the NRG boreholes is illustrated for thermal/mechanical units for low temperatures (<100 °C) in Table 3-13. Variations for high temperatures (>100 °C) are shown in Table 3-14. The data were compiled in this manner rather than for each 25 °C interval because thermal conductivity is not strongly temperature dependent. No data were available for tuff rocks in the UO (undifferentiated overburden) unit, or for an upper portion of the TCw unit (Tpcrv, Tpcrn, Tpcpul, and Tpcmn). Additional thermal conductivity data from the TSw2 unit (Tptpmn) from the Single Heater Test block is shown in Table 3-15. These specimens were all tested in the air-dried state, that is, in the as-received condition. Results are consistent with results from the specimens from NRG

boreholes. Thermal conductivity data from the TSw2 unit in the Drift Scale Test block is presented in Table 3-16. The specimens were all tested in the saturated state, and results are also consistent with specimens from NRG boreholes.

Thermal conductivities are lower for dried specimens and highest for saturated specimens. Thermal conductivities, averaged over all boreholes, ranged, depending on temperature and saturation state, from 1.2 to 1.9 W/mK for TCw, from 0.4 to 0.9 W/mK for PTn, from 1.0 to 1.7 W/mK for TSw1, and from 1.5 to 2.3 W/mK for TSw2. The data show distinct differences between the nonwelded tuffs of the PTn and the welded tuffs of the TCw, TSw1, and TSw2 units. PTn consistently shows the lowest conductivities, while the TCw and TSw2 units have the highest values. TSw1 specimens span a larger range of thermal conductivities and are intermediate in value.

Evaluation of the mean values (Tables 3-13, 3-14, 3-15, and 3-16) indicates that thermal conductivity was affected by saturation and, to a lesser degree, temperature. Thermal conductivity generally increased with increasing saturation

Table 3-11. Mineral Abundances for 20 Samples from the Drift Scale Test Block of the Exploratory Studies Facility

Sample #	Quartz	Cristobalite	Albite	Sanidine	Tridymite	Zeolite Phase
SDM-MPBX1-1.0-1.2-D	12	23	34	32	nd[2]	nd[2]
SDM-MPBX1-21.0-21.2-D	10	27	25	35	3	nd[2]
SDM-MPBX1-21.0-21.2-D dup	9	27	25	35	4	nd[2]
SDM-MPBX1-31.9-32.1-D	4	31	32	33	nd[2]	nd[2]
SDM-MPBX1-40.4-40.6-D	20	14	31	29	7	nd[2]
SDM-MPBX1-62.0-D	11	26	27	36	nd[2]	nd[2]
SDM-MPBX1-80.7-80.9-D	12	24	31	33	nd[2]	nd[2]
SDM-MPBX2-29.0-29.2-D	4	31	33	32	nd[2]	nd[2]
SDM-MPBX2-29.0-29.2-D dup	4	29	34	33	nd[2]	nd[2]
SDM-MPBX2-48.6-D	18	18	30	34	nd[2]	nd[2]
SDM-MPBX2-72.0-D	6	28	34	32	nd[2]	nd[2]
SDM-MPBX2-85.0-D	7	26	34	33	nd[2]	nd[2]
SDM-MPBX3-17.5-17.7-D	8	25	33	34	nd[2]	nd[2]
SDM-MPBX3-38.5-38.7-D	5	31	24	40	nd[2]	nd[2]
SDM-MPBX3-85.6-D	11	22	30	34	nd[2]	3 - stilbite
AOD-HDFR#1-9.0-D	13	20	32	34	nd[2]	2 -clinoptilolite
AOD-HDFR#1-9.0-D dup	15	20	31	35	nd[2]	Trace - clinoptilolite
AOD-HDFR#1-48.5-D	12	25	26	37	nd[2]	nd[2]
AOD-HDFR#1-68.6-D	12	25	24	39	nd[2]	nd[2]
AOD-HDFR#1-98.0-D	37	4	29	30	nd[2]	nd[2]

[1] Mineral abundances were normalized to 100% after subtracting out the estimated abundance of the corundum standard. Sums may differ from 100% due to roundoff error.

[2] nd - not detected.

Table 3-12. Rock Thermal Conductivities at Temperatures Below 100 °C

Thermal/ Mechan ical Unit	Thermal Conductivity (W/mK)											
	Saturated			Partially Saturated			Air Dry			Dry		
	Sample Mean	Sample Standard Deviation	Sample Count	Sample Mean	Sample Standard Deviation	Sample Count	Sample Mean	Sample Standard Deviation	Sample Count	Sample Mean	Sample Standard Deviation	Sample Count
TCw	1.89	0.12	18	1.39	0.56	18	1.58	0.16	9	1.17	0.35	18
PTn	0.92	0.13	42	0.57	0.12	33	0.35	0.13	12	0.38	0.10	49
TSw1	1.70	0.19	50	1.23	0.46	11	1.21	0.12	30	0.98	0.26	59
TSw2	2.29	0.42	51	ND	ND	ND	1.66	0.10	24	1.49	0.44	48

Source: CRWMS M&O (1997d)

ND—No data. Sample refers to the number of test measurements, not the number of specimens tested. Measurements were made at multiple temperatures and during both heating and cooling for some specimens.

Table 3-13. Rock Thermal Conductivities at Temperatures Above 100 °C

Thermal/Mechanical Unit	Thermal Conductivity (W/mK)		
	Dry		
	Sample Mean	Sample Standard Deviation	Sample Count
TCw	1.53	0.17	57
PTn	0.42	0.14	102
TSw1	1.15	0.15	173
TSw2	1.59	0.10	125

Source: CRWMS M&O (1997d)

a) All high temperature data were acquired above 100 °C. 'Sample' refers to the number of test measurements, not the number of specimens tested. Measurements were made at multiple temperatures and during both heating and cooling for some specimens.

Table 3-14. Thermal Data for Specimens from the Single Heater Test Block (Air Dry)

Apparatus[a]	Temperature	Thermal Conductivity (W/m-K)				Mean[b]	Standard Deviation[b]
		ESF-H1-0.6-B	ESF-H1-11.3-B	ESF-H1-11.6-B	ESF-H1-19.9-B		
LT	30	1.50	1.76	1.37	1.76	1.60	0.195
LT	50	1.52	1.79	1.40	1.77	1.62	0.1913113
LT	70	1.54	1.81	1.44	1.77	1.64	0.178699
TCA	70	1.58	1.96	1.61	1.89	1.76	0.1930458
TCA	110	1.56	1.88	1.57	1.80	1.70	0.1621471
TCA	155	1.61	1.85	1.62	1.79	1.72	0.1209339
TCA	200	1.60	1.82	1.59	1.78	1.70	0.1195478
TCA	245	1.57	1.81	1.60	1.75	1.68	0.1158663
TCA	289	1.48	1.76	1.56	1.69	1.62	0.1260622
Mean		1.55	1.83	1.53	1.78	N/A	N/A
Standard Deviation		0.04	0.06	0.10	0.05	N/A	N/A

Source: CRWMS M&O 1996b

(a) The low temperature (LT) was used for testing at 70 °C and below; the TCA (Thermal Conductivity Analyzer) was used for 70 °C and above.
(b) Mean and standard deviation for all data in table are 1.67 W/(m-K) and 0.15 W/m-K), respectively.

Table 3-15. Thermal Conductivity Data for Specimens from the DST Block (Saturated)

Distance from Collar (ft)	Max. Temp. (°C)	Thermal Conductivity (W/m-K)					
		30 °C	50 °C	70 °C	Mean	STD[a]	N[a]
ESF-SDM-MPBX1-C							
1.0	70	2.2	2.2	2.2	2.2	0.0	3
32.1	70	2.2	2.2	2.1	2.2	0.0	3
40.6	70	2.1	2.1	2.1	2.1	0.0	3
62.0	70	2.2	2.2	2.2	2.2	0.0	3
80.5	70	2.15	2.1	2.1	2.1	0.0	3
	N[a] =	5	5	5			
	Mean =	2.2	2.2	2.1			
	STD[a] =	0.1	0.0	0.0			
ESF-SDM-MPBX2-C							
13.0	70	2.1	2.1	2.1	2.1	0.0	3
29.0	70	2.1	2.1	2.1	2.1	0.0	3
48.4	70	2.0	2.0	2.0	2.0	0.0	3
71.5	70	2.3	2.3	2.3	2.3	0.0	3
84.6	70	2.0	2.0	2.1	2.0	0.1	3
	N[a] =	5	5	5			
	Mean =	2.1	2.1	2.1			
	STD[a] =	0.1	0.1	0.1			
ESF-SDM-MPBX3-C							
3.0	70	1.9	1.9	1.9	1.9	0.0	3
17.7	70	2.1	2.1	2.1	2.1	0.0	3
38.7	70	2.2	2.2	2.2	2.2	0.0	3
72.0	70	2.2	2.2	2.2	2.2	0.0	3
85.3	70	2.1	2.1	2.1	2.1	0.0	3
	N[a] =	5	5	5			
	Mean =	2.1	2.1	2.1			
	STD[a] =	0.1	0.1	0.1			
ESF-AOD-HDFR1-C							
8.6	70	2.0	2.0	2.0	2.0	0.0	3
32.2	70	2.1	2.1	2.1	2.1	0.0	3
48.7	70	1.9	1.9	2.0	1.9	0.0	3
68.8	70	2.1	2.1	2.0	2.1	0.0	3
97.5	70	2.3	2.3	2.3	2.3	0.0	3
	N[a] =	5	5	5			
	Mean =	2.1	2.1	2.1			
	STD[a] =	0.2	0.1	0.1			
All Drift Scale Characterization Boreholes							
N[a] =		20	20	20			
Mean =		2.1	2.1	2.1			
STD[a] =		0.1	0.1	0.1			
All Specimens, All Temperatures							
N[a] =		60					
Mean =		2.1					
STD[a] =		0.1					

[a]N = Number of samples; STD = standard deviation.

NOTE: Air dried. Lithostratigraphic unit: Tptpmn, except for HDFR1-97.5-C, which may be from Tptpll.

Table 3-16. Mean Coefficient of Thermal Expansion (CTE) During Heat-Up

T/M Unit	Satura-tion State	Statistics[a]	Mean CTE on Heat-Up (10⁻⁶/°C)										
			25-50°C	50-75°C	75-100°C	100-125°C	125-150°C	150-175°C	175-200°C	200-225°C	225-250°C	250-275°C	275-300°C
TCw	Saturated	Mean	7.09	7.62	8.08	10.34	13.17	15.20	16.99	18.99	21.38	27.42	42.99
		Std. Dev.	0.43	0.15	0.50	1.52	1.23	1.57	1.41	0.96	1.23	1.94	37.35
		Count	4	4	4	4	4	4	4	3	3	3	3
	Dry	Mean	6.60	8.29	9.62	10.53	12.69	14.90	17.03	20.68	29.64	36.49	49.15
		Std. Dev.	1.49	0.99	1.06	1.60	1.55	1.91	2.31	5.41	21.88	16.97	34.24
		Count	10	10	10	7	7	7	7	7	7	7	7
			25-50°C	50-75°C	75-100°C	100-125°C	125-150°C	150-175°C	175-200°C	200-225°C	225-250°C	250-275°C	275-300°C
PTn	Saturated	Mean	4.46	4.28	-1.45	-30.42	5.54	4.47	0.64	-4.65	-9.79	-13.46	-12.96
		Std. Dev.	0.38	1.61	3.63	21.47	0.41	0.79	1.03	4.05	7.85	11.12	12.90
		Count	4	4	4	4	3	3	3	2	2	2	2
	Dry	Mean	4.55	4.24	3.36	-4.78	6.46	5.69	3.61	0.56	-2.98	-5.81	-7.25
		Std. Dev.	0.74	1.46	2.40	11.12	0.98	1.41	2.58	5.81	9.12	11.36	10.80
		Count	12	12	12	10	10	10	10	10	10	10	10
			25-50°C	50-75°C	75-100°C	100-125°C	125-150°C	150-175°C	175-200°C	200-225°C	225-250°C	250-275°C	275-300°C
TSw1	Saturated	Mean	6.56	7.32	6.83	6.92	10.72	14.28	20.98	36.82	41.64	42.76	43.81
		Std. Dev.	1.16	0.60	1.60	3.28	1.74	3.26	7.01	20.49	17.35	13.19	13.65
		Count	10	10	10	10	10	9	9	8	8	8	8
	Dry	Mean	6.29	7.60	8.39	8.96	10.37	15.51	23.67	34.24	34.00	36.07	38.74
		Std. Dev.	1.22	1.02	0.89	1.20	1.38	4.53	11.07	20.30	13.70	13.23	13.78
		Count	33	33	33	28	28	27	26	25	25	25	25
			25-50°C	50-75°C	75-100°C	100-125°C	125-150°C	150-175°C	175-200°C	200-225°C	225-250°C	250-275°C	275-300°C
TSw2	Saturated	Mean	7.14	7.47	7.46	9.07	9.98	11.74	13.09	15.47	19.03	25.28	37.19
		Std. Dev.	0.65	1.51	1.21	2.41	0.77	1.28	1.40	1.75	3.09	6.87	14.27
		Count	19	19	19	19	19	19	19	16	16	16	16
	Dry	Mean	6.67	8.31	8.87	9.37	10.10	10.96	12.22	14.52	20.79	25.13	35.13
		Std. Dev.	1.20	0.42	0.40	0.55	0.88	1.16	1.50	2.57	17.03	10.07	14.56
		Count	40	40	40	40	40	38	38	35	35	35	35

NOTE: [a]Std.Dev.—Standard Deviation.   Source: CRWMS M&O 1997d

NOTE: The negative coefficients of the thermal expansion are apparent and indicate that some samples were observed to contract in the temperature ranges containing the negative values. The contractions are interpreted as drying phenomena.

and temperature. Sharp increases in thermal conductivity are observed near 100 °C for several oven-dried specimens. These increases are as yet unexplained, but may be associated with a change in instrumentation at 100 °C or with the vaporization of remaining water. For Exploratory Studies Facility Single Heater Test block specimens, thermal conductivity appeared to increase sharply at 70 °C. This response was probably associated with the change in instrumentation at that temperature. At temperatures above 100 °C, thermal conductivity shows little temperature dependence. Decreases in conductivity with increasing temperature observed in saturated specimens are attributed to dehydration (Brodsky et al., 1997).

The effective thermal diffusivity (thermal conductivity divided by the product of density and specific heat) of crushed tuff was measured in two bench-scale tests. The first test ran ~502 h, and the second, 237 h. In each test, a cylindrical volume (1.58 m³) was filled with crushed tuff particles ranging in size from 12.5 mm to 37.5 mm (0.5 in to 1.5 in) to form an effective porosity of 0.48. Heat was generated by an axial heater. Temperatures near the heater reached 700 °C, with a significant volume of material exceeding 100 °C. Thermal diffusivity was estimated post-test using three different analysis methods. Estimates of thermal diffusivity were 5.0×10^{-7} m²/s to 6.6×10^{-7} m²/s, of the same order of magnitude as the thermal diffusivity used for crushed backfill in the Total System Performance Assessment 1993 (CRWMS M&O, 1997d).

Attempts have been made to correlate thermal conductivity with an easily measured physical property such as porosity, as discussed in Rautman and McKenna (1997). The relationship between thermal conductivity and porosity, saturation, and temperature is discussed by Lappin et al. (1982), who documented thermal conductivities of the major silicate phases in tuff and also discussed calculation of matrix conductivities from conductivities of components and measured values of porosity and saturation using the Woodside and Messmer (1961) geometric mean approach (Nimick, 1989). Nimick (1990) introduced use of the Brailsford and Major (1964) equation for calculating matrix thermal conductivity to replace the geometric mean equation used previously by Lappin et al. (1982), and used the Brailsford and Major (1964) equation for calculation of matrix porosity. Nimick (1990) summarized data for 15 samples of Topopah Spring Tuff and estimated matrix conductivities and in situ thermal conductivities for the TSw1 and TSw2 thermal/mechanical units.

The porosities of samples adjacent to many of the NRG borehole thermal conductivity test specimens (from the same original piece of core) have been determined. Both the geometric mean equation and the Brailsford and Major (1964) equations were investigated, and based on the measured saturation and porosity of each specimen, the matrix conductivity for each specimen was calculated using each model. These matrix conductivities were then averaged for each model to obtain conductivity at zero porosity. The change in conductivity with increasing porosity was then calculated for each model (CRWMS M&O, 1997d). Unfortunately, lithologies also change with increasing porosity, and so it is difficult to isolate the effects of changing one variable.

An alternative method of viewing the conductivity versus porosity relationship is reported in Brodsky et al. (1997).

Thermal Expansion

Thermal expansion is the tendency of a material to undergo a nearly proportional degree of volume or length change caused by a change in temperature. The coefficient of thermal expansion, is usually recorded as a change in strain (linear dimension per unit original length) per °C. The relationship presented by Weast (1974) is:

$$l_t = l_0 \left(1 + \alpha T \right) , \tag{3-3}$$

where l_0 = length at 0 °C, l_t = length at T °C, and α = coefficient of linear thermal expansion.

Thermal expansion measurements on tuffs correlative with those found in the Exploratory Studies Facility are reported in Lappin (1980) for samples from U#25A#1, Well J-13, and G-tunnel; and in Schwartz and Chocas (1992) for 109 specimens from UE-25A#1, USW G-1, USW G-2, USW G-4, and USW GU-3. Seventy-eight of the 109 specimens reported by Schwartz, and Chocas (1992) were tested unconfined, and 31 were tested at a nominal 10 MPa confining pressure.

Thermal expansion measurements were made using a push rod dilatometer. Test specimens were right circular cylinders, ~50.8 mm in length and 25.4 mm in diameter. Moisture contents were either air dry (as received), oven dry, or vacuum saturated. Tests were conducted at ambient pressure over the temperature range of 30 °C to over 300 °C. Temperature was ramped at 1 °C per minute. Specimens under saturated conditions were tested up to 100 °C. The test procedure is given in SNL TP-203, *Measurement of Thermal Expansion of Geologic Supplies Using a Push Rod Dilatometer.* The closest applicable ASTM standard is ASTM E228–85, *Standard Test Method for Linear Thermal Expansion of Solid Materials With a Vitreous Silica Dilatometer.* This standard describes equipment similar to that used in SNL TP-203. An additional applicable ASTM standard is ASTM D4535–85, *Standard Test Methods for Measurement of Thermal Expansion of Rock Using a Dilatometer.*

Temperature and displacement data were obtained throughout a heating and cooling cycle to determine the coefficient of thermal expansion. Coefficients of thermal expansion change with temperature, and so calculations were performed over specified temperature intervals. Mean thermal expansion coefficients were calculated over 25 °C intervals from data obtained at the end points of the interval. Instantaneous coefficients of thermal expansion were calculated from linear least square fits to data contained within 5 °C windows spaced 25 °C apart. The instantaneous coefficient of thermal expansion data are given in Brodsky et al. (1997).

Mean coefficients of thermal expansion from surface-based borehole specimens are presented by thermal/mechanical unit and saturation in Table 3-17 for heating phases and in Table 3-18 for cooling phases. Qualified thermal expansion data only exist

Table 3-17. Mean Coefficient of Thermal Expansion During Cool-Down for Borehole Samples

T/M Unit	Saturation State	Statistics	Mean Coefficient of Thermal Expansion on Cool-Down ($10^{-6}/°C$)										
			300-275°C	275-250°C	250-225°C	225-200°C	200-175°C	175-150°C	150-125°C	125-100°C	100-75°C	75-50°C	50-35°C
TCw	Saturated	Mean	14.72	21.97	33.53	37.01	23.81	18.48	15.72	13.51	12.09	10.78	10.85
		Std. Dev.	3.76	6.79	16.44	26.18	10.01	3.25	1.96	1.48	1.28	1.36	1.96
		Count	3	3	3	3	4	4	4	4	4	4	4
	Dry	Mean	17.46	26.34	36.95	33.72	22.86	17.58	13.89	11.77	10.21	9.35	6.59
		Std. Dev.	3.70	6.88	11.50	14.21	3.16	2.00	2.39	2.22	1.56	1.15	2.20
		Count	7	7	7	7	7	7	7	7	9	9	9
			300-275°C	275-250°C	250-225°C	225-200°C	200-175°C	175-150°C	150-125°C	125-100°C	100-75°C	75-50°C	50-35°C
PTn	Saturated	Mean	15.58	9.12	7.20	6.39	6.98	6.29	5.93	5.36	5.12	4.33	1.94
		Std. Dev.	1.04	0.84	0.29	0.17	1.51	0.78	0.47	0.36	0.34	0.84	2.93
		Count	2	2	2	2	3	3	3	4	4	4	4
	Dry	Mean	11.22	7.91	6.78	6.45	6.47	6.53	6.11	5.80	5.52	4.82	2.41
		Std. Dev	2.46	1.00	0.81	0.90	1.14	1.35	1.30	1.16	0.89	0.84	0.86
		Count	10	10	10	10	10	10	10	10	12	12	12
			300-275°C	275-250°C	250-225°C	225-200°C	200-175°C	175-150°C	150-125°C	125-100°C	100-75°C	75-50°C	50-35°C
TSw1	Saturated	Mean	15.07	19.87	24.05	26.15	27.57	26.66	28.19	19.89	11.46	9.92	9.35
		Std. Dev.	4.68	7.82	9.85	7.37	8.36	9.91	18.04	8.05	2.01	1.54	1.06
		Count	8	8	8	8	8	8	10	10	10	10	10
	Dry	Mean	16.68	20.71	24.16	23.26	26.74	25.34	25.55	17.78	10.53	9.22	6.95
		St. Dev.	4.13	7.78	10.61	7.29	9.32	9.35	14.94	8.53	2.15	1.51	2.49
		Count	25	25	25	25	26	27	28	28	33	33	33
			300-275°C	275-250°C	250-225°C	225-200°C	200-175°C	175-150°C	150-125°C	125-100°C	100-75°C	75-50°C	50-35°C
TSw2	Saturated	Mean	21.89	27.83	26.55	21.38	17.31	14.06	12.49	11.52	10.27	9.48	8.81
		Std. Dev	6.16	10.36	10.01	5.70	3.07	1.38	1.32	2.00	0.62	0.63	0.62
		Count	16	16	16	16	19	19	19	19	19	19	19
	Dry	Mean	20.57	24.31	24.20	21.16	18.45	14.34	11.74	10.51	9.54	8.87	7.48
		Std. Dev	4.88	7.55	8.08	6.24	9.36	4.23	3.03	2.26	1.79	1.56	1.99
		Count	35	35	35	35	38	38	40	40	40	40	40

Source: CRWMS M&O (1997d, Table 5-18, p. 5-59)

NOTE: T/M = thermal-mechanical; Std. Dev. = standard deviation

Table 3-18. Mean Coefficient of Thermal Expansion for TSw2 Specimens from the Single Heater Test Block

Temperature Range (°C)		Mean Coefficient of Thermal Expansion (MCTE) (10^-6/°C)										
Low	High	ESF-H1-0.6-A	ESF-H1-11.3-A	ESF-H1-11.6-A	ESF-H1-22.2-A	ESF-MPBX-3-12.5-A	ESF-MPBX-4-5.0-A	ESF-MPBX-4-18.3-A	ESF-BJ-1-1.0-A	ESF-BJ-1-10.0-A	Mean MCTE	Standard Deviation
25	50	6.69	7.59	7.21	7.62	7.61	8.27	7.34	7.30	7.63	7.47	0.42
50	75	8.39	8.65	8.71	8.90	8.47	9.39	9.23	8.81	9.38	8.88	0.38
75	100	8.88	9.48	9.47	9.61	8.88	10.30	10.00	9.80	10.30	9.64	0.53
100	125	8.77	9.83	9.82	10.1	9.1	11.2	10.6	10.1	10.6	10.01	0.76
125	150	9.37	10.6	10.5	10.5	9.47	12.3	11.6	10.7	11.4	10.72	0.95
150	175	10.6	11.0	11.1	10.9	10.0	12.5	11.9	11.3	12.0	11.26	0.77
175	200	16.7	11.5	12.5	12.0	10.9	13.3	12.6	12.2	13.3	12.78	1.66
200	225	26.3	13.3	15.2	13.3	13.5	16.3	14.8	13.9	16.1	15.86	4.08
225	250	25.8	17.4	21.1	17.5	15.5	23.1	19.0	17.1	18.7	19.47	3.28
250	275	29.8	25.7	35.0	27.1	29.1	39.0	28.8	24.9	27.8	29.69	4.54
275	300	46.2	44.6	58.5	59.3	62.1	61.1	50.8	40.3	42.4	51.70	8.65
300	275	14.3	27.0	26.6	30.1	28.2	29.1	28.3	27.2	27.6	26.49	4.70
275	250	33.3	33.9	32.9	42.4	42.4	41.5	35.7	30.2	32.6	36.10	4.73
250	225	30.4	30.7	32.0	38.8	38.2	40.1	30.6	25.3	28.8	32.77	5.08
225	200	24.1	21.7	31.1	25.6	21.7	26.7	21.7	19.1	22.2	23.77	3.59
200	175	23.2	16.6	27.2	18.6	16.0	19.3	16.9	15.9	17.1	18.98	3.83
175	150	12.3	13.7	17.3	14.8	13.1	15.0	13.9	13.6	14.2	14.21	1.42
150	125	8.7	11.9	13.6	12.9	11.9	13.5	12.1	11.8	12.3	12.08	1.43
125	100	7.60	10.9	11.6	11.8	10.8	11.7	11.1	10.7	11.2	10.82	1.27
100	75	7.23	10.3	10.9	11.2	10.2	11.1	10.6	10.3	10.6	10.27	1.19
75	50	6.22	9.13	9.56	9.97	9.08	10.30	9.60	9.20	9.54	9.18	1.18
50	35	5.67	8.48	8.79	9.26	8.41	9.30	8.55	8.55	8.86	8.43	1.08

Source: CRWMS M&O 1996b

for the TCw, PTn, TSw1, and TSw2 thermal/mechanical units. The mean thermal expansion coefficient does show some borehole-to-borehole variation, which is obscured by the data averaging in these tables. The mean thermal expansion coefficient was highly temperature dependent, and ranged, depending upon temperature and saturation state, from $6.6 \times 10^{-6}/°C$ to $50 \times 10^{-6}/°C$ for TCw, from negative values to $16 \times 10^{-6}/°C$ for PTn, from $6.3 \times 10^{-6}/°C$ to $44 \times 10^{-6}/°C$ for TSw1, and from $6.7 \times 10^{-6}/°C$ to $37 \times 10^{-6}/°C$ for TSw2.

Additional data for TSw2 thermal/mechanical unit (Tptpmn) samples from the Single Heater Test region of the Exploratory Studies Facility Thermal Test Facility are presented in Table 3-19. The mean thermal expansion coefficients for these samples ranged from 7.5 to $52 \times 10^{-6}/°C$, and as shown, were temperature dependent during the heating cycle (CRWMS M&O, 1996b).

Statistical summaries for mean coefficients of thermal expansion for specimens from the Drift Scale Test block are given in Table 3-20 for heating and cooling cycles (CRWMS M&O (1997a). Single Heater Test and Drift Scale Test values are generally consistent. There is more variability in the Drift Scale Test data, but that study included a much larger sampling volume. Drift Scale Test mean values are consistently higher than those for the Single Heater Test and are generally higher during cooling. However, the mean Single Heater Test values are typically within one standard deviation of the mean Drift Scale Test values.

At a "transition temperature" of 150° to 200 °C, the mean thermal expansion coefficient increases more steeply for the welded tuff but decreases for the nonwelded tuff (Tables 3-17–3-19). A transition is expected in the welded devitrified specimens caused by phase changes in tridymite and cristobalite. These minerals occur, with or without quartz, as primary devitrification products in many samples of Yucca Mountain welded tuffs. Phase transitions in synthetic tridymite occur at ~117° and 163 °C, and in synthetic cristobalite at ~272 °C (Papike and Cameron, 1976), and involve notable changes in volume. Phase transition temperatures have been shown to vary significantly due to lattice variations found in natural occurrences of these minerals, which are usually mixed-phase material (Thompson and Wennemer, 1979). Previous and current work on the mineralogy of welded tuff from TCw, TSw1, TSw2, and TSw3 suggests that these mixed-phase assemblages are dominant. Hysteresis is associated with the phase changes because the phases invert at higher temperature during heating than during cooling (Brodsky et al., 1997).

Some specimens that displayed sensitivity to transition temperature were analyzed to assess the role of the maximum test temperature. Specimens from approximately the same depth and same borehole were tested to different temperatures. The results showed that as long as the maximum test temperature remained below the transition temperature, the specimens did not permanently change dimension (Brodsky et al., 1997).

The sharp increase in mean coefficients of thermal expansion beginning at ~200 °C in welded tuffs is not attributable to thermally induced fracturing or differential expansion, since these behaviors would not be significant during the second heating phase, and the Drift Scale Test data indicate sharp increases for both cycles. Three specimens from the Drift Scale Test suite of tests did not show the increase in mean coefficients of thermal expansion at elevated temperature. Two of these specimens appeared to initiate phase changes below 200 °C, and one specimen appeared to undergo essentially no phase change. This difference in behavior is attributed to different concentrations of cristobalite and tridymite

Thermal expansion was independent of saturation state for welded specimens but did depend on saturation state for the nonwelded specimens. Nonwelded specimens with high moisture contents contracted during testing near 100 °C, causing a temporary sharp decrease in mean thermal expansion coefficient at ~100 °C. This is presumed to be due to reduction of pore water and dehydration of hydrated glass. However, there is insufficient information to determine a mechanism for the shrinkage. The expansion characteristics of the welded specimens, on the other hand, seemed to be independent of saturation state, and the curves for different saturation states are similar (Brodsky et al., 1997).

A suite of confined thermal expansion tests was conducted to determine if strain hysteresis and transition temperature effects would be suppressed by elevated confining pressures. These test results are discussed in detail in Martin et al. (1997). Confining pressure effects for specimens tested between 1 and 30 MPa were very small (Martin et al., 1997), and so data from these pressures were averaged together and compared with data from unconfined tests. At temperatures below ~150 °C, the coefficient of thermal expansion is slightly lower for unconfined tests than for confined tests. However, at higher temperatures (150–250 °C), the mean values approach one another, and the scatter among the unconfined tests encompasses the results for the confined tests. Over the higher temperature range, coefficients of thermal expansion therefore appear to be independent of confining pressure.

The effect of specimen size was investigated in a study using samples from Busted Butte and from boreholes USW G-1, USW G-2, USW G-3, and USW G-4. For all depths except the Calico Hills (Tac) unit sampled in USW G-2, mean thermal expansion coefficients are higher for small specimens than for large specimens. However, the difference in mean values is always within the error in the measurement or within one standard deviation of the mean, and data were insufficient to formulate conclusions about size effect.

The correlation between thermal properties and mineralogy was very poor but may be explained in several ways. First, the data set is very limited. Second, the mineralogical analyses were not performed on the test specimens themselves, but on pieces of rock taken from near the test specimens. Third, hysteresis in the thermal expansion data may be influenced by other rock properties besides mineralogy, such as lithophysae content, and a larger database and further analyses would be required to separate the influences of each of these variables (Brodsky et al., 1997).

Table 3-19. Mean Coefficient of Thermal Expansion for Drift Scale Test Characterization Borehole Samples (TSw2; Air Dry)

First Heating Cycle

MCTE on Heat-Up (10^{-6}/°C)

	25-50°C	50-75°C	75-100°C	100-125°C	125-150°C	150-175°C	175-200°C	200-225°C	225-250°C	250-275°C	275-300°C	300-325°C
N[a] =	17	17	17	17	17	17	17	17	17	17	17	17
Mean =	7.34	8.99	9.73	10.22	10.91	12.20	14.74	22.31	27.34	33.88	54.13	52.28
STD[a] =	0.57	0.47	0.54	0.58	0.79	1.04	4.79	18.09	15.70	6.94	12.18	13.42
95%[a] =	0.27	0.22	0.26	0.28	0.38	0.49	2.28	8.60	7.46	3.30	5.79	7.29

First Cooling Cycle

Mean CTE on Cool-Down (10^{-6}/°C)

	325-300°C	300-275°C	275-250°C	250-225°C	225-200°C	200-175°C	175-150°C	150-125°C	125-100°C	100-75°C	75-50°C	50-30°C
N[a] =	13	17	17	17	17	17	17	17	17	17	16	15
Mean =	15.74	24.07	35.63	36.01	26.50	24.19	18.30	14.14	12.36	11.05	10.24	9.67
STD[a] =	1.88	5.70	8.39	8.32	4.69	9.82	7.37	2.61	1.76	0.84	1.24	0.66
95%[a] =	1.02	2.71	3.99	3.96	2.23	4.67	3.50	1.24	0.84	0.40	0.61	0.33

Second Heating

MCTE on Heat-Up (10^{-6}/°C)

	25-50°C	50-75°C	75-100°C	100-125°C	125-150°C	150-175°C	175-200°C	200-225°C	225-250°C	250-275°C	275-300°C	300-325°C
N[a] =	17	17	17	17	17	17	17	17	17	17	17	17
Mean =	7.22	8.87	9.63	10.24	11.28	13.22	19.37	22.66	24.82	37.84	46.78	34.46
STD[a] =	0.76	0.59	0.54	0.61	0.98	2.98	14.97	10.66	5.47	9.52	11.63	10.17
95%[a] =	0.36	0.28	0.26	0.29	0.46	1.42	7.12	5.07	2.60	4.52	5.53	4.84

Second Cooling

MCTE on Cool-Down (10^{-6}/°C)

	325-300°C	300-275°C	275-250°C	250-225°C	225-200°C	200-175°C	175-150°C	150-125°C	125-100°C	100-75°C	75-50°C	50-30°C
N[a] =	17	17	17	17	17	17	17	17	17	17	17	17
Mean =	16.50	26.48	37.06	36.39	26.44	23.64	17.46	13.75	12.01	11.36	10.16	9.81
STD[a] =	2.23	4.89	8.42	7.94	4.51	9.46	6.15	2.08	1.18	1.83	0.65	0.70
95%[a] =	1.06	2.32	4.00	3.77	2.14	4.50	2.92	0.99	0.56	0.87	0.31	0.33

Table 3-20. Thermal Capacitance (Cp) for TSw1 and TSw2 Thermal/Mechanical Units

TSw1				TSw2			
Temperature (°C)	Mean Cp	Standard Deviation (J cm^{-3} K^{-1})	No. Of Tests	Temperature (°C)	Mean Cp	Standard Deviation (J cm^{-3} K^{-1})	No. Of Tests
25	1.58	0.05	3	25	1.79	0.11	7
50	1.68	0.05	3	50	1.88	0.11	7
75	1.80	0.05	3	75	1.97	0.11	7
100	1.91	0.05	3	100	2.16	0.11	7
125	2.03	0.06	3	125	2.32	0.11	7
150	2.14	0.11	3	150	2.45	0.13	7
175	2.13	0.10	3	175	2.43	0.18	7
200	2.09	0.07	3	200	2.40	0.16	7
225	2.07	0.06	3	225	2.39	0.17	7
250	2.05	0.05	3	250	2.39	0.19	7
275	2.03	0.05	3	275	2.39	0.22	7
300	2.03	0.06	3	300	2.43	0.26	7

Source: CRWMS M&O (1997d)

Heat Capacity

Heat capacity is the amount of heat required to change the temperature of a substance by a given amount. It is defined (Halliday and Resnick, 1974) as:

$$H = \frac{\Delta J}{\Delta T}, \qquad (3\text{-}4)$$

where ΔJ = quantity of heat (J) and ΔT = change in temperature (K).

The database for heat capacity measurements consists of theoretical values calculated by Nimick and Connolly (1991) from chemical and mineralogical data, and experimental values reported in Brodsky et al. (1997).

Bulk chemical analyses of 20 tuff samples from Yucca Mountain were used to calculate heat capacities of the solid components of the tuffs as a function of temperature. The data were combined with grain density, matrix porosity, lithophysal-cavity abundance, mineral abundance, in situ saturation, and the properties of water to estimate rock mass thermal capacitances. Calculations were completed for thermal/mechanical units (TCw, PTn, TSw1, TSw2, TSw3, CHn2V, CHn1z, and CHn2z) over the temperature range of 25° to 275 °C. Summary mineralogical and chemical data are reported in Connolly and Nimick (1990), and thermal capacitance calculations and results are given in Nimick and Connolly (1991). Data for TSw1 and TSw2 are presented later in this section in comparison with experimentally determined values.

Heat capacity was measured for 10 air-dried specimens from UE-25 NRG-4 and UE-25 NRG-5. The test method and test results are described and analyzed in detail in Brodsky et al. (1997).

Test specimens were air-dried cylinders, ~57.0 mm in length and 50.8 mm in diameter. Tests were conducted over the temperature range of 30° to 300 °C using an adiabatic pulse calorimeter. This instrument applies a known quantity of electrical energy to a specimen and measures the resulting rise in specimen temperature. The test procedure is presented in SNL-TP-204, *Measurement of Specific Heat of Geologic Samples by Adiabatic Pulse Calorimetry*. The closest applicable ASTM procedure is ASTM D4611, *Standard Practice for Specific Heat of Rock and Soil*. This standard is based on the drop in calorimetry and provides less accurate results than the adiabatic calorimeter method given in TP-204.

Thermal capacitance, which is heat capacity multiplied by specimen density, is summarized in Table 3-20. A complete data presentation is included in Brodsky et al. (1997) and CRWMS M&O (1997d). Thermal capacitance is higher for TSw2 than for TSw1. Mean thermal capacitance ranges from 1.6 J cm^{-3} K^{-1} to 2.1 J cm^{-3} K^{-1} for TSw1 and from 1.8 J cm^{-3} K^{-1} to 2.5 J cm^{-3} K^{-1} for TSw2 (Table 3-20).

Experimentally determined values of heat capacity increased with temperature, reaching a localized peak of 2.4 and 2.1 cm^{-3} K^{-1} at ~150 °C to 170 °C for the TSw1 and TSw2 units, respectively. This peak may be related to a phase change. However, the data presented here were insufficient to correlate peaks more specifically. The peaks in specific heat at these temperatures do occur at a temperature range associated with the phase change in tridymite (163 °C). It also is evident that there were no significant changes in measured specific heat for these air-dried specimens at 100 °C, indicating that dehydration effects were minor (Brodsky et al., 1997).

Values of thermal capacitance calculated, along with the experimental data in CRWMS M&O (1997d) are shown in Table 3-20. For both the theoretical and experimental data, values for TSw2 are higher than for TSw1. The theoretical calculations show a decrease at 100 °C related to the heat of vaporization of water. The test specimens were air dried and showed no comparable decrease. The two sets of data roughly coincide. However, it should be noted that without the decrease at 100 °C in the calculated data, the discrepancy between measured and calculated values for TSw1 would be maintained.

Mechanical Properties

A comprehensive series of mechanical property measurements was conducted on specimens prepared from borehole cores, which included elastic constants, strength, and deformation moduli for specimens from all thermal/mechanical units. One objective of the measurements was to establish a baseline set of properties, to study the vertical and lateral variability of bulk and mechanical properties at Yucca Mountain. Boyd et al. (1996a) noted there is little lateral variability among the boreholes located along the axis of the North Ramp. However, there is significant vertical variability related to large differences in lithology, and smaller differences in smaller-scale fabric and pore structure characteristics. Tests were performed on specimens of tuff without fractures and large lithophysae (>~1 cm).

Static and Dynamic Elastic Constants

Young's modulus and Poisson's ratio are the primary mechanical deformation indices of rock and are indicators of the elastic response of rock to stress. Static Young's modulus and Poisson's ratios were computed from the stress-strain data obtained for the specimens tested in confined and unconfined compression. In addition, dynamic elastic moduli were computed from compressional- and shear-wave velocities measured under ambient laboratory conditions.

The test procedure for compressional- and shear-wave velocity measurements involved measurement of ultrasonic compressional- and shear-wave velocities both parallel and normal to the core axis of the test specimen. One compressional and two orthogonally polarized shear waves were measured parallel to the axis. One compressional and one polarized shear wave with a vibration direction parallel to the core axis were measured in the radial direction. These data were used to compute the elastic anisotropy of the specimen. In addition, the compressional- and shear-wave velocity, combined with specimen density, were used to compute dynamic Young's modulus and Poisson's ratios (Martin et al., 1994).

In general, the Young's modulus of the tuff depends on the degree of welding. Nonwelded tuff is weak and exhibits low

Young's moduli (Tables 3-21 and 3-22). In contrast, the welded tuffs are stronger and exhibit significantly greater Young's moduli. In borehole NRG-6, for example, the moduli range from <1 GPa for the nonwelded units to near 40 GPa for the welded units. The greatest moduli are observed for specimens recovered from units TCw and TSw2. The Young's moduli observed on specimens from TSw1 are somewhat lower than those for the other welded units. The standard deviation in the Young's moduli for each thermal/mechanical unit is large. Specimens separated by very small vertical distances, having nominally the same texture and composition, exhibited large changes in moduli (CRWMS M&O, 1997d).

For specimens from the TSw2 thermal/mechanical unit in the Single Heater Test block of the Exploratory Studies Facility Thermal Test Facility, elastic constants calculated from unconfined compression tests were fairly consistent, with a mean Young's modulus of 32.4 GPa and a mean Poisson's ratio of 0.17 (CRWMS M&O, 1996b).

Elastic constants calculated from unconfined compression tests on specimens from the TSw2 thermal/mechanical unit (Tptpmn lithostratigraphic unit) in the Drift Scale Test block were slightly higher than values from the Single Heater Test block, but generally were consistent with values from NRG borehole samples. The mean Young's modulus was 36.8 GPa, and the mean Poisson's ratio, 0.201 (CRWMS M&O, 1997f).

Dynamic Young's moduli were computed from the velocity and density data. Dynamic moduli exceed static moduli. In general, the ratio of dynamic to static moduli is less than two; the difference is greatest at lower moduli, and decreases as the moduli increase.

Compressional- and shear-wave velocities were measured in the dry and in the saturated condition on each specimen tested in unconfined and confined compression (Table 3-23). The compressional- and shear-wave velocities are greatest in the TCw and TSw2 welded units, and lowest in the nonwelded PTn units (CRWMS M&O, 1997d). The compressional-wave velocity increases with saturation, and the shear-wave velocity decreases. These effects are consistent with theoretical models of seismic waves in porous media.

The elastic anisotropies of specimens tested under unconfined and confined compression were calculated from the ultrasonic velocity data. The anisotropy ranges from <5% to 15%. Of the densely welded units, the TSw1 unit exhibited the largest anisotropy, probably caused by the presence of oriented lithophysae and vapor-phase altered zones.

Compressive Strength

Compressive strength of a rock is its ability to withstand compressive stress without failure. Compressive strength of intact rock is measured in the laboratory by subjecting a cylindrical test piece to a compressive load parallel to its axis until it fails. Compressive strength is the maximum stress at failure and is computed from the maximum load and the cross-sectional area of the test piece. Confined (or triaxial) compressive strength is

Table 3-21. Mean and Standard Deviation of Intact Rock Elastic Modulus for Thermal/ Mechanical Units and Lithostratigraphic Units

T/M Unit	Lithostratigraphic Unit	Elastic (Young's) Modulus*	
		Mean (GPa)	Std. Dev. (GPa)
UO		3.86	1.86
	Tmr	3.36	2.69
	Tpki	4.28	0.81
	Tpbt5	-	-
TCw		29.36	10.80
	Tpcrv	-	-
	Tpcrn	15.16	8.39
	Tpcpul	23.72	6.89
	Tpcpmn	34.97	7.44
	Tpcpll	34.31	5.24
	Tpcpln	34.28	7.66
PTn		2.54	4.02
	Tpcpv	7.18	7.61
	Tpbt4	1.82	1.10
	Tpy	5.24	3.05
	Tpbt3	0.25	0.07
	Tpp	1.05	0.60
	Tpbt2	0.83	0.60
	Tptrv	0.93	0.87
TSw1		20.36	6.75
	Tptrn	20.72	6.25
	Tptrl	10.45	3.27
	Tptpul	21.44	8.54
TSw2		33.03	5.94
	Tptpmn	32.93	5.47
	Tptpll	27.54	7.49
	Tptpln	35.48	5.45
TSw3			
	Tptpv	37.43	15.02
CHn		5.63	1.55
	Tpbt1	3.90	-
	Tac	6.50	0.57

*Quasi-static
Std. Dev.—standard deviation
- No data

Source: CRWMS M&O (1997d)

determined by subjecting the cylindrical test specimen to a uniform lateral confining pressure in addition to the axial load.

Unconfined compressive strength. Unconfined compressive strength of a wide range of tuff samples from the Yucca Mountain area have been measured and reported by a number of workers (Olsson and Jones, 1980; Olsson, 1982; Price and Jones, 1982; Price and Nimick, 1982; Price et al., 1982a, 1982b, 1984; Price, 1983; Nimick et al., 1985). Compressive strengths of the various units of the Paintbrush Group from the USW G-1 to G-4 borehole series are given in Price et al. (1985).

For unconfined compression tests, described in Brechtel et al. (1995) and CRWMS M&O (1997d), cylindrical tuff specimens were tested to failure at a constant strain rate of 10^{-5} s^{-1} under ambient temperature and pressure conditions. Nominal specimen size was 101.6 mm in length and 50.8 mm in diameter. The specimens were saturated with distilled water. Testing was accomplished following the procedures in ASTM D3148, *Standard Test Method for Elastic Moduli of Intact Rock Core Specimens in Uniaxial Compression*, and International Society of Rock Mechanics procedure, *Suggested Methods for Determining*

Table 3-22. Mean and Standard Deviation of Intact Rock Poisson's Ratio for Thermal/Mechanical Units and Lithostratigraphic Units

T/M Unit	Lithostratigraphic Unit	Poisson's Ratio*	
		Mean	Std. Dev.
UO		0.09	0.07
	Tmr	0.03	0.03
	Tpki	0.14	0.05
	Tpbt5	-	-
TCw		0.21	0.11
	Tpcrv	-	-
	Tpcrn	0.20	0.03
	Tpcpul	0.19	0.02
	Tpcpmn	0.27	0.29
	Tpcpll	0.21	0.05
	Tpcpln	0.21	0.02
PTn		0.20	0.06
	Tpcpv	0.11	0.06
	Tpbt4	0.16	0.01
	Tpy	0.16	0.01
	Tpbt3	0.24	0.06
	Tpp	0.29	0.12
	Tpbt2	0.23	0.09
	Tptrv	0.21	0.08
TSw1		0.23	0.07
	Tptrn	0.23	0.06
	Tptrl	0.29	0.02
	Tptpul	0.25	0.13
TSw2		0.21	0.04
	Tptpmn	0.21	0.03
	Tptpll	0.21	0.06
	Tptpln	0.24	0.05
TSw3		0.24	0.13
	Tptpv	0.24	0.13
CHn		0.17	0.12
	Tpbt1	0.11	-
	Tac	0.20	0.16

* Quasi-static
Std. Dev.—standard deviation
- No data

Source: CRWMS M&O (1997d)

Table 3-23. Comparison of the Mean and Standard Deviation of Compressional and Shear Wave Velocities on Dry Specimens—NRG and SD Boreholes

T/M Unit	TCw	TCw	PTn	PTn	TSw1	TSw1	TSw2	TSw2
Wave Type	P	S	P	S	P	S	P	S
Mean Velocity, (km s^{-1})	4.369	2.811	2.687	1.900	3.739	2.352	4.158	2.786
Standard Deviation	0.494	0.132	0.703	0.308	0.393	0.204	0.672	0.110
Range (Min.–Max.)	2.865 — 4.787	2.352 — 3.009	1.538 — 3.750	1.365 — 2.464	2.767 — 4.629	2.048 — 2.928	2.673 — 4.709	2.408 — 3.007
Number of Samples	66	42	11	11	57	38	50	47

Source: CRWMS M&O (1997d)

Uniaxial Compressive Strength and Deformability of Rock Materials (Brown, 1981). Static Young's modulus and Poisson's ratios were computed by performing linear least squares fits to the stress and strain data collected between 10% and 50% of the fracture strength. The reported fracture strength was the maximum stress exerted on the specimen.

Results of unconfined compression tests on specimens from NRG and SD boreholes indicate that the unconfined compressive strengths vary depending on the welding, porosity, and fabric of the rock. Welded tuffs exhibited higher strengths than nonwelded tuffs. Within the welded units, the variations in strengths are related to the presence and size of lithophysae and vapor-phase altered zones (Table 3-24).

The fracture modes in most of the unconfined compression tests were very similar. In most cases, the fractures terminated in the specimen end caps, and there was no evidence of shear cone development.

Specimens from TCw typically exhibit the greatest strengths; strengths in excess of 300 MPa are observed for this unit. In contrast, the weakest specimens are from the PTn unit, with strengths generally <10 MPa. Large variability is observed for TSw1 and TSw2. The strengths for these units vary from 25 to 250 MPa and show no consistent trend between strength and depth.

In most cases, very little inelastic volumetric strain (dilatancy) is observed in the welded TCw, TSw1, and TSw2 tuffs (CRWMS M&O, 1997d). Many crystalline rocks begin to dilate at stresses as low as 50% of the fracture strength. However, in the welded tuffs, little nonlinearity in the volumetric strain is observed until the specimens are very near failure. The manner in which cracks grow and interact in tuff appears different from that observed in other crystalline rocks in which microcrack porosity is dominant. The data suggest that axial cracks extend without interacting with other cracks until failure is imminent. Similar effects also have been reported by Brace et al. (1966) and Scholz (1968).

Results of unconfined compression tests on TSw2 show a large scatter in strengths similar to that observed in other testing of Yucca Mountain tuffs (Brechtel et al., 1995; CRWMS M&O, 1997d). Unconfined compressive strengths range from 75.1 to 243.8 MPa, with a mean of 143.2 MPa and a standard deviation of ±50.3 MPa (CRWMS M&O, 1996b). Moisture contents for these specimens were not controlled, and differences in moisture contents may have contributed to the scatter in strengths. The mode of failure for all specimens is dominated by brittle axial cracking. There is no correlation between strength and mode of failure, and all of the specimens failed explosively (CRWMS M&O, 1996b).

Results of unconfined compression tests on 16 samples from the TSw2 thermal/mechanical unit (Tptpmn lithostratigraphic unit) also show a large scatter in strengths (CRWMS M&O, 1997f). Strengths range from 71.3 to 324.1 MPa, with a mean value of 176.4 MPa and a standard deviation of ±66 MPa. The highest and lowest strengths were obtained on specimens that were only 4 m apart—neither of these specimens had notable surface features that might predict anomalous behavior.

Confined compressive strength. Confined compression experiments were carried out on cores with a length to diameter ratio of two and nominal diameters of 25.4 or 50.8 mm (Brechtel et al., 1995; CRWMS M&O, 1997d). All specimens were tested saturated, in a drained condition. The general procedure for testing the specimens was the same as for the unconfined compression tests, except that the specimens were jacketed in copper and deformed in a pressure vessel at a fixed confining pressure. The procedure used for the measurements generally conformed to International Society of Rock Mechanics, *Suggested Methods for Determining the Strength of Rock Materials in Triaxial Compression* (Brown, 1981) and ASTM D2664–86, *Standard Test Method for Triaxial Compressive Strength of Undrained Rock Core Specimens Without Pore Pressure Measurements.* Measurements were performed at confining pressures of 5 and 10 MPa, and pressure was held constant to ±0.1 MPa. The instrumented specimens were monotonically loaded to failure at a nominal strain rate of 10^{-5} s^{-1}. Young's modulus and Poisson's ratios were computed between 10% and 50% of the stress difference at failure. Results of confined compression tests indicate that the axial stress difference at failure increases with increasing confining pressure (Table 3-25). The specimens tested in confined compression failed in shear; that is, the fractures formed on shear places with little evidence of axial splitting. In most cases, a visibly evident shear plane developed. However, there was no evidence of conjugate fracture sets forming in any of the specimens.

To examine temperature effects, 17 confined compression experiments were also performed at a nominal temperature of 150 °C on borehole specimens from thermal/mechanical unit TSw2. Measurements were conducted at effective confining pressures of 1, 5, and 10 MPa, with a pore pressure of 5 MPa. The experiments were performed at a nominal strain rate of 10^{-6} s^{-1}. The confining pressure was held constant to ±0.25 MPa, and the pore pressure was maintained constant at 5.00 ± 0.25 MPa. The elevated temperature was generated by means of band heaters positioned on the outside of the pressure vessel. The temperature was monitored inside the pressure vessel with a thermocouple positioned at the midpoint of the specimen.

The high-temperature, confined compression test data indicate a clear increase in strength between 1 MPa and 5 MPa (CRWMS M&O, 1997d). However, there is no apparent increase in the mean strength of the tuff between 5 and 10 MPa confining pressure. Comparison of room and elevated-temperature (150 °C) tests suggests that the effect of temperature on the strength of welded tuff from thermal/mechanical units TSw2 is small. Similarly, Young's modulus and Poisson's ratios measured at elevated temperatures (150 °C) are not significantly different from those measured at room temperature.

The heterogeneity in the tuff poses a problem in analyzing the effect of pressure on the strength in terms of Mohr-Coulomb criteria. As seen with the unconfined compression test data, the strengths may vary by as much as a factor of two, even over a limited depth interval. Without a suite of nominally identical specimens, a

Table 3-24. Mean and Standard Deviation of Uniaxial Compressive Strength for Thermal Mechanical and Lithostratigraphic Units

T/M Unit	Lithostratigraphic Unit	Uniaxial Compressive Strength Mean (MPa)	Std. Dev. (MPa)
UO		7.14	4.92
	Tmr	7.60	7.48
	Tpki	6.75	1.84
	Tpbt5	-	-
TCw		127.54	92.66
	Tpcrv	-	-
	Tpcrn	36.02	32.98
	Tpcpul	62.14	26.54
	Tpcpmn	167.19	64.12
	Tpcpll	230.94	105.30
	Tpcpln	172.33	58.09
PTn		6.37	6.97
	Tpcpv	16.61	19.81
	Tpbt4	3.54	1.74
	Tpy	19.30	6.99
	Tpbt3	1.30	0.71
	Tpp	3.41	2.29
	Tpbt2	3.14	2.05
	Tptrv	3.50	2.02
TSw1		58.22	30.69
	Tptrn	58.99	28.27
	Tptrl	26.73	9.25
	Tptpul	65.21	44.40
TSw2		167.90	65.53
	Tptpmn	187.49	64.92
	Tptpll	103.82	60.68
	Tptpln	144.44	39.57
TSw3		16.40	-
	Tptpv	16.40	-
CHn		23.23	3.04
	Tpbt1	20.50	-
	Tac	24.60	2.69

Std. Dev.—standard deviation

\- No data

Source: CRWMS M&O (1997d)

Table 3-25. **Strength Parameters Calculated from the Confined Compressive Test Results, Stratigraphic Units**

Thermal/ Mechanical Unit	Lithostratigraphic Unit	Cohesion (MPa)	Angle of Internal Friction (degrees)	Data Source for Strength Parameters			
				Core Hole	Sample ID	Confined Pressure (MPa)	Axial Stress (MPa)
TCw	Tiva Canyon Tuff: Crystal-poor Middle Nonlithophysal Zone (Tpcpmn)	14.2	65	NRG-3	263.3-E	5	158.2
				NRG-3	263.3-D	10	291.4
				NRG-3	265.7-B	5	295.9
				NRG-3	265.7-D	5	275.1
				NRG-3	265.7-F	5	180.3
				NRG-3	265.7-C	10	369.0
				NRG-3	265.7-E	10	339.1
				NRG-3	265.7-G	10	309.9
TCw	Tiva Canyon Tuff: Crystal-poor Lower Lithophysal Zone (Tpcpll)	50.6	54	NRG-6	22.2-B	10	439.7
				NRG-6	22.2-C	10	434.6
				NRG-6	22.2-D	10	343.4
				NRG-6	22.2-E	10	417.3
				NRG-6	22.2-F	5	395.3
				NRG-6	22.2-G	5	391.2
				NRG-6	22.2-H	5	396.8
				NRG-6	22.2-I	5	307.1
				NRG-6	22.2-J	0	315.8
				NRG-6	22.2-K	0	284.2
				NRG-6	22.2-L	0	313.8
				NRG-6	22.2-M	0	332.4
TCw	Tiva Canyon Tuff: Crystal-poor Lower Nonlithophysal Zone (Tpcpln)	27.8	63	SD-12	144.7-A	0	278.4
				SD-12	156.0-A	5	313.6
				SD-12	161.7-A	10	418.2
				SD-12	165.8-A	0	169.8
				SD-12	176.1-A	10	410.4
				SD-12	195.3-A	0	259.9
				SD-12	202.9-A	5	323.3
TSw1	Topopah Spring Tuff: Crystal-rich Nonlithophysal Zone (Tptrn)	6.4	55	NRG-4	527.0-A	5	76.5
				NRG-4	527.0-B	5	108.9
				NRG-4	527.0-D	5	113.6
				NRG-4	527.0-E	5	90.0
				NRG-4	527.0-G	5	67.5
				NRG-4	527.0-C	10	160.8
				NRG-4	527.0-F	10	142.1
				NRG-4	527.0-H	10	131.4
				NRG-4	527.0-I	10	133.0
TSw1	Topopah Spring Tuff: Crystal-rich Nonlithophysal Zone (Tptrn)	3.4	53	NRG-6	416.0-B	10	119.9
				NRG-6	416.0-C	10	147.9
				NRG-6	416.0-E	10	61.1
				NRG-6	416.0-F	5	77.1
				NRG-6	416.0-G	5	39.6
				NRG-6	416.0-H	5	84.7
				NRG-6	416.0-I	5	58.8

Table 3-25.   Strength Parameters Calculated from the Confined Compressive Test Results, Stratigraphic Units (Continued)

Thermal/ Mechanical Unit	Lithostratigraphic Unit	Cohesion (MPa)	Angle of Internal Friction (degrees)	Data Source for Strength Parameters			
				Core Hole	Sample ID	Confined Pressure (MPa)	Axial Stress (MPa)
TSw1	Topopah Spring Tuff: Crystal-rich Nonlithophysal Zone (Tptrn)	15.1	51	NRG-7/7A	344.4-A	10	185.0
				NRG-7/7A	344.4-B	10	139.1
				NRG-7/7A	344.4-D	10	208.3
				NRG-7/7A	344.4-E	5	56.5
				NRG-7/7A	344.4-F	5	163.9
				NRG-7/7A	344.4-G	5	75.9
				NRG-7/7A	344.4-H	5	110.2
				NRG-7/7A	344.4-I	0	106.2
				NRG-7/7A	344.4-J	0	89.5
				NRG-7/7A	344.4-K	0	97.3
				NRG-7/7A	344.4-L	0	88.7
TSw1	Topopah Spring Tuff: Crystal-rich Nonlithophysal Zone (Tptrn)	9.9	48	SD-12	339.4-A	5	88.8
				SD-12	343.4-A	10	124.6
				SD-12	354.1-A	0	44.4
				SD-12	370.1-A	5	87.4
				SD-12	401.5-A	10	102.9
				SD-12	406.1-A	0	48.4
				SD-12	410.2-A	5	99.4
TSw2	Topopah Spring Tuff: Crystal-poor Middle Nonlitho-physal Zone (Tptpmn)	42.8	50	NRG-7/7A	865.4-A	10	325.2
				NRG-7/7A	865.4-B	10	354.0
				NRG-7/7A	865.4-D	10	235.5
				NRG-7/7A	865.4-E	10	316.7
				NRG-7/7A	865.4-C	5	259.8
				NRG-7/7A	865.4-F	5	322.3
				NRG-7/7A	865.4-G	5	255.1
				NRG-7/7A	865.4-H	5	231.6
				NRG-7/7A	865.4-I	0	215.8
				NRG-7/7A	865.4-J	0	232.0
				NRG-7/7A	865.4-K	0	239.1
				NRG-7/7A	865.4-L	0	248.5
TSw2	Topopah Spring Tuff: Crystal-poor Middle Nonlitho-physal Zone (Tptpmn)	36.9	46	SD-9	761.5-A	0	231.5
				SD-9	764.8-A	6	231.9
				SD-9	764.8-B	10	218.3
				SD-9	766.0-A	15	210.8
				SD-9	768.7-A	0	254.5
				SD-9	771.7-A	0	160.8
				SD-9	774.6-B	0	60.1
				SD-9	774.6-C	10	212.3
				SD-9	775.8-A	15	280.9
				SD-9	775.8-B	6	181.7
				SD-9	815.9-B	10	344.2
				SD-9	817.1-A	15	334.4
				SD-9	826.7-A	0	224.9
				SD-9	832.8-C	0	183.3

Table 3-25.   Strength Parameters Calculated from the Confined Compressive Test Results, Stratigraphic Units (Continued)

Thermal/ Mechanical Unit	Lithostratigraphic Unit	Cohesion (MPa)	Angle of Internal Friction (degrees)	Data Source for Strength Parameters			
				Core Hole	Sample ID	Confined Pressure (MPa)	Axial Stress (MPa)
TSw2	Topopah Spring Tuff: Crystal-poor Lower Nonlithophysal Zone (Tptpln)	22.7	58	SD-12	1073.3-B	5	216.1
				SD-12	1077.1-B	10	246.1
				SD-12	1107.1-B	0	162.0
				SD-12	1112.1-B	5	280.6
				SD-12	1118.9-B	10	288.9
				SD-12	1209.0-B	0	128.0
TSw3	Topopah Spring Tuff: Crystal-poor Vitric Zone (Tptpv)	3.5	47	SD-12	1279.7-A	5	52.2
				SD-12	1284.2-A	10	80.0
				SD-12	1299.9-A	0	16.4

Source: CRWMS M&O (1997d)

Mohr-Coulomb failure envelope is difficult to establish. Data are judged to be incomplete for the PTn and CHn1 units.

Tensile Strength

Indirect tensile strength tests, commonly referred to as Brazilian tests, were carried out using a procedure adhering to ASTM D3967, *Standard Test Method for Splitting Tensile Strength of Intact Rock Core Specimens.* The test is simple in principle. A load is applied to a cylindrical specimen with its axis normal to the loading direction. Tensile stress develops in the center of the cylinder. The force is increased until the specimen fails by an extension fracture along the loading plane. The tensile strength is computed from the force at failure.

Indirect tensile strength tests were performed on specimens from boreholes that were 38.1 mm in length and 50.8 mm in diameter (Table 3-26). Tensile strengths generally range between 0.2 and 16 MPa. The weakest specimens are from the nonwelded PTn thermal/mechanical unit. The greatest strengths are observed in the TCw welded tuff. In general, TSw1 is weaker than TSw2.

Shear Strength

Mathematical simulation of the response of rock to mining and drilling requires the use of failure criteria for the rock. One commonly used criteria is the Mohr-Coulomb criterion, which defines the limiting state of stress for static equilibrium with the material at which inelastic deformation begins (Jaeger and Cook, 1979). The criterion itself is expressed as follows:

$$\tau = C_0 + \sigma \tan \phi , \qquad (3\text{-}5)$$

where τ = shear stress on the failure plane at the onset of failure, σ = normal stress on the failure plane at the onset of failure, C_0 = cohesion, and ϕ = angle of internal friction

The shear stress at the onset of failure is also the shear strength of rock, which is defined by its two components, cohesion (C_0)

and angle of internal friction (ϕ). Results of unconfined compressive tests and triaxial (confined) compression tests can be used to determine τ and ϕ. As noted in the discussion of compression test results, because of the heterogeneity in the tuff, it is difficult to establish a Mohr-Coulomb failure envelope without a suite of nominally identical specimens.

Time-Dependent (Creep) Behavior

If a solid is subjected to a load (stress) within its elastic limit, it instantaneously experiences an amount of deformation (strain), which disappears on the removal of the load. If the load is maintained at the same level, the solid will continue to deform beyond the instantaneous deformation at a slow rate depending on the level of the applied stress. This continuing deformation with time in spite of no increase in stress is referred to as time-dependent deformation or creep deformation.

Seven creep experiments were performed on right circular cylinders of TSw2 tuff (CRWMS M&O, 1997d). The specimens had a length-to-diameter ratio of two and a specimen diameter of 50.8 mm. The experiments were performed at a constant confining pressure of 10 MPa and a temperature of 225 °C. The procedure used for these measurements was based on ASTM D4406, *Standard Test Method for Creep of Cylindrical Rock Core Specimens in Triaxial Compression.* Measurements were carried out in a compact creep apparatus, described by Martin et al. (1995). The key features of the system included independent controls for the confining pressure, pore pressure, temperature, and axial force producing the differential stress on the sample. Loading continued under constant temperature, differential stress, and confining pressure until the sample failed or the experiment was terminated.

Creep measurements were made on specimens from the TSw2 thermal/mechanical unit (Tptpmn lithostratigraphic unit) at nominal differential stresses of 40, 70, 100, and 130 MPa, at a fixed confining pressure of 10 MPa, and at a temperature of 225 °C (Table 3-27). The duration of the experiments ranged from

Table 3-26. Brazilian Tensile Strength for Thermal/Mechanical and Lithostratigraphic Units

T/M Unit	Lithostratigraphic Unit	Brazilian Tensile Strength	
		Mean (MPa)	Standard Deviation (MPa)
UO		0.88	0.72
	Tmr	1.23	1.27
	Tpki	0.70	0.24
	Tpbt5	-	-
TCw		8.88	3.66
	Tpcrv	-	-
	Tpcrn	5.23	2.80
	Tpcpul	9.63	0.73
	Tpcpmn	10.10	4.75
	Tpcpll	12.20	4.39
	Tpcpln	11.00	1.75
PTn		0.66	1.17
	Tpcpv	2.64	2.04
	Tpbt4	0.3	-
	Tpy	3.00	-
	Tpbt3	0.10	-
	Tpp	0.19	0.19
	Tpbt2	0.35	0.31
	Tptrv	0.1	-
TSw1		5.48	2.32
	Tptrn	5.49	2.09
	Tptrl	4.81	1.36
	Tptpul	5.69	3.06
TSw2		8.91	3.39
	Tptpmn	11.56	3.80
	Tptpll	8.29	2.99
	Tptpln	7.75	2.59
TSw3		3.97	0.32
	Tptpv	3.97	0.32
CHn		-	-
	Tpbt1	-	-
	Tac	-	-

- No Data

Source: CRWMS M&O 1997d

Table 3-27. Creep Measurements, Borehole USW NRG-7/7A

Depth (ft):	776.6	807.6	808.3	858.4	1264.5	1281.4	1400.5
Thermal/Mechanical Unit:	TSw2	TSw2	TSw2	TSw2	TSw2	TSw2	TSw2
Date Test initiated:	3/28/95	3/28/95	2/17/95	3/28/95	9/30/93	9/30/93	9/30/93
As Tested Bulk Density (g/cm³):	2.249	2.273	2.295	2.307	2.295	2.295	2.295
Average Grain Density (g/cm³):	2.536	2.534	2.526	2.526	2.591	2.592	2.515
Porosity (%):	11.3	10.3	9.2	8.7	11.4	11.5	8.8
P Velocity (km/s):	4.357	4.426	4.455	4.488	4.064	4.431	N/A
S1 Velocity (km/s):	2.774	2.775	2.800	2.816	2.563	N/A	N/A
S2 Velocity (km/s):	2.730	2.712	2.848	2.769	2.510	N/A	N/A
Radial P Velocity (km/s):	4.388	4.418	4.471	4.465	4.450	4.618	N/A
Radial S Velocity (km/s):	2.808	2.836	2.824	2.847	2.703	2.815	N/A
Temperature (°C):	225	225	225	225	225	225	225
Confining Pressure (MPa)	10	10	10	10	10	10	10
Stress Difference (MPa)	70	40	129	100	98	132	131
Strain @ 1,000s (millistrain)	2.34	1.08	3.02	2.69	3.01	3.70	3.47
Strain termination (millistrain)	2.52	1.16	3.06	2.88	3.25	4.04	3.84
Duration of Test (days)	43.5	43.5	68.3	43.5	29.5	29.5	29.5
Duration of Test (millions of seconds)	3.76	3.76	5.90	3.76	2.55	2.55	2.55

Source: CRWMS M&O (1997d)

NOTES: 1. Nominal Sample Dimensions: Length = 101.60 mm; Diameter = 50.80 mm.
2. P is the compressional wave, S1 and S2 are the two orthogonally polarized shear waves.

2.55×10^6 s to 5×10^6 s (30–68 d). At higher stress differences, the data show very small increases in the axial strain. The experiments conducted at stress differences between 40 and 100 MPa show smaller strain accumulations, and for the test conducted at a differential stress of 40 MPa, no strain accumulation is observed. Each test was terminated before failure of the specimen.

Hardness

Schmidt hammer rebound hardness measurements were conducted on samples from core holes to produce early strength estimates and to supplement the rock mechanics test data. The measurements were performed following International Society of Rock Mechanics suggested methods (Brown, 1981), and the analysis of the results incorporates suggested improvements to the International Society of Rock Mechanics methods by Goktan and Ayday (1993). Pieces of core were selected on nominal 3 m (10 ft) intervals down hole and clamped in a testing anvil weighing a minimum of 20 kg (44.1 lbs). Rebound hardness measurements were conducted on a group of 20 samples (Table 3-28).

Correlations and Parametric Effects for Mechanical Properties

Anisotropy

In general, rock properties often depend on the direction of measurement relative to such characteristics as bedding and fabric. Elastic anisotropy of specimens tested in confined and unconfined compression was calculated from ultrasonic velocity data. The anisotropy ranges from <5% to 15%. The TSw1 thermal/mechanical unit exhibited the largest anisotropy, probably because of the presence of oriented lithophysae and vapor-phase altered zones. Price (1983) found that dynamic elastic moduli for samples of the densely welded Topopah Spring Member showed that anisotropy of elastic properties for orientations parallel and perpendicular to the rock fabric was insignificant.

Lithophysae

The units of the Topopah Spring Tuff do contain varying amounts of lithophysae. Lithophysae are generally spherical to sometimes slightly flattened cavities up to several tens of centimeters in diameter. The lithophysal cavities often have on their inner wall a thin layer of feldspars, silica minerals, and other vapor-phase mineral deposits. The functional porosity, which is the total of lithophysal pore volume, nonlithophysal pore volume, and volume of clay minerals, has been related by Price and Bauer (1985) to Young's modulus and unconfined compressive strength.

Porosity

The vertical variability in the elastic and strength properties of the tuffs at Yucca Mountain is large. Even considering the data within thermal/mechanical units, the scatter for each property is large. Variations of a factor of two or more are common. Price and Bauer (1985) sorted the elastic and fracture properties of tuff

according to porosity in an attempt to compare specimens with similar properties. They observed a good correlation between modulus and porosity. However, there is still significant scatter in the moduli even for similar samples with nearly identical porosities. In spite of the apparent correlation of modulus with porosity for porosities between 8% and 60%, it is still difficult to predict modulus based on porosity data (CRWMS M&O, 1997d).

A similar correlation exists between ultrasonic wave velocities and porosity. Based on data collected on NRG and SD specimens, compressional- and shear-wave velocities decrease with increasing porosity (CRWMS M&O, 1997d). However, as with Young's modulus, at each porosity, there is large scatter in the measured velocities.

The distribution of strengths observed in unconfined compression tests similarly correlates with porosity (CRWMS M&O, 1997d). As with the elastic properties, the strength decreases with increasing porosity, and there is substantial scatter at each porosity. Even within a very narrow range of porosities, and presumably similar composition and distribution of lithophysae and vapor-phase altered zones, the variability in strength is large. Most of this scatter has been attributed to differences in pore distribution (Price et al., 1993).

Temperature Effects

The effect of temperature on the TSw2 welded tuff produced somewhat conflicting results (CRWMS M&O, 1997d). The application of moderate confining pressure (5 or 10 MPa) resulted in a significant increase in the stress difference at failure. However, comparison of data collected at ambient temperature with those at elevated temperature at very low confining pressures indicates that the effect of temperature on strength is small. More data are needed to assess the influence of temperature on strength.

Effect of Sampling on Parametric Studies

The current data set does not adequately isolate parametric effects, such as temperature, pressure, and saturation, because of the large variability in properties between lithologic units and within lithologic units (CRWMS M&O, 1997d). Samples were selected to characterize vertical variability within the thermal/mechanical units. Comparisons were made with samples from different depths in the boreholes, and attempts were made to group properties to define how temperature and confining pressure influence behavior in a specific rock unit. However, variability between samples may be greater than the parametric effect being investigated.

Scaling Intact Properties

Scaling of baseline data on intact rock properties for site and time may be important for modeling approaches that do not assume an equivalent continuum. In this case, the intact properties of rock blocks between fractures represent an intermediate size between the rock mass and the laboratory sample. Scaling laws need to be developed based on the requirements of design and performance assessment modeling approaches.

Table 3-28. Summary of Statistical Data for Schmidt Hammer Tests in Thermomechanical Units—NRG Holes

Core Hole	UO (Tuff "X")			All TCw			TCw-NL			TCw-LR			PTn			All TSw1			TSw1-NL			TSw1-lr			TSw2		
	mean	sdev	N	mean	sdev	N	mean	sdev	N	mean	sdev	N	mean	sdev	N	mean	sdev	N	mean	sdev	N	mean	sdev	N	mean	sdev	N
UE25 NRG-2				45.94	3.11	13	47.68	1.92	4	45.16	3.31	9															
UE25 NRG-2A				44.09	2.60	10	44.09	2.60	10																		
UE25 NRG-2B				27.60	11.61	11	24.61	10.62	9	41.09	1.04	2															
UE25 NRG-3				39.76	9.17	21	34.67	11.98	10	44.39	3.49	11															
UE25 NRG-4													17.88	2.86	2	46.21	4.88	16	46.53	5.06	14	43.99	3.59	2	42.70	9.25	5
UE25 NRG-5																11.83	—	1				11.83	—	1			
UE25 NRG-6				49.71	4.89	9	47.50	2.84	3	50.81	5.53	6	38.70	7.96	2	46.53	7.06	31	48.34	4.38	16	44.61	8.86	15	48.02	9.32	10
UE25 NRG-7/7A				41.08	7.99	6	41.08	7.99	6				26.07	11.77	4	48.35	4.28	19	48.37	4.68	10	48.33	4.06	9	48.29	5.10	14
ALL HOLES																											
—Range				11.48 - 56.95			11.48 - 50.15			36.33 - 56.95			15.85 - 44.33			11.83 - 56.58			35.75 - 56.58			11.83 - 54.70			23.65 - 56.45		
—Mean				41.01			37.82			45.78			27.18			46.45			47.71			44.59			47.23		
—Sdev				10.10			11.47			4.65			11.53			7.23			4.66			9.70			7.53		
—N				70			42			28			8			67			40			27			29		

Sdev = standard deviation
N = number of samples
NL = Nonlithophysal
LR = Lithophysae Rich

Two important parameters for scaling intact properties are the volume of the rock and the time duration of the applied stress at elevated temperature. Price (1986) studied the effect of sample size on the unconfined compressive strength and elastic moduli of welded tuff in a study conducted on TSw2 welded specimens recovered from Busted Butte, adjacent to Yucca Mountain. The data show a decrease of more than a factor of two in compressive strength as the specimen diameter increases from 25.4 mm to 228.6 mm.

The data pertaining to time dependence reveal an inconsistency. The study of strain rate dependence for Busted Butte specimens shows that subcritical crack propagation substantially affects failure strengths. However, failure strengths for specimens at room temperature were not different from those at 150 °C. Subcritical crack propagation is a thermally activated process, and so, if strain rate dependence is observed, it is expected that some temperature dependence also would be observed.

Mechanical Properties of Fractures

Discontinuities, such as joints, bedding planes, faults, and fractures, cause the properties of a rock mass in situ to be different from those of intact rock tested in the laboratory. In general, strength and deformational properties of the rock mass will be lower than those measured in the laboratory. An understanding of the mechanical properties of discontinuities is necessary to evaluate their effect on rock mass properties.

Most rock masses contain natural fractures that are called joints if they show little or no visible offset. The presence of these features in the rock mass has important effects on the overall thermal/mechanical and hydro/mechanical response of the rock mass. They can increase the compliance, reduce the strength, alter the thermal conductivity, and act as pathways for fluid movement. Because of their potential importance in design and performance assessment, fracture properties such as normal stiffness, shear stiffness, cohesion, and coefficient of friction are important.

Natural fractures from boreholes were tested for stiffness and strength (Olsson and Brown, 1994, 1995, 1997). These studies used the rotary shear technique for measuring fracture stiffness and shear strength. This technique has been used for at least 20 yr and has certain advantages over other configurations for shear testing. Details on implementation of this test technique,

along with discussions of its relative advantages and disadvantages and many important results, can be found in a number of papers (Biegel et al., 1992; Christensen et al., 1974; Kutter, 1974; Olsson, 1987, 1988; Olsson and Brown, 1993; Xu and de Frietas, 1988; Tullis and Weeks, 1986; Weeks and Tullis, 1985; Yoshioka and Scholz, 1989).

In the rotary shear tests, the specimens were composed of two short, hollow tubes of rock that were divided at mid-height by the fracture to be tested. Specimens were prepared from the drill core by subcoring perpendicular to the fracture. Outer specimen diameters ranged from 44.3 to 82.0 mm (1.74–3.22 in). Test specimens were potted into metal specimen holders with gypsum cement. The metal specimen holders were then bolted into the load-frame. Stress was applied normal to the fracture while the change in fracture aperture was measured with linear variable differential transformers. Torque was then applied to cause sliding on the interface, and shear displacements and shear strength were measured. All samples were sheared at constant normal stress at room temperature and in the air-dry condition (Olsson, 1987; Olsson and Brown, 1997).

The normal stiffness for Yucca Mountain fractures increases with increasing normal stress, as is typical for interfaces of any type (Olsson and Brown, 1994, 1995). However, there is wide variability from fracture to fracture, which is indicated by the large standard deviations (Table 3-29). The coefficient of friction seems to be slightly greater for TSw2 than for TSw1, but the one standard deviation band includes essentially all the other values.

There are too few data to draw quantitative conclusions about fracture strengths for TCw and CHn1. For TSw1 and TSw2, where there are more data, the friction angle, expressed as tan ϕ, seems to be slightly greater for TSw2 than for TSw1, 0.87 ($\phi = 41°$) ± 0.09 versus 0.77 ($\phi = 38°$) ± 0.08, respectively. The roughness characteristics of the fracture surfaces agree qualitatively with the simple mathematical model of Brown (1995) derived from fracture data in many other rock types (Tables 3-30 and 3-31).

ROCK MASS PROPERTIES

Analyses to support the design of the repository are required to address the potential impacts of seismic, thermal, and mechanical loading. These analyses require knowledge of rock properties

Table 3-29.   **Average Fracture Normal Stiffness at 2.5 MPa Normal Stress for Each of the Four Thermal/Mechanical Units Sampled**

Thermal/Mechanical Unit	Normal Stiffness at 2.5 MPa[a] (MPa/mm)	Number of Tests
TCw	37.3 ± 4.0	2
TSw1	50.1 ± 17.6	16
TSw2	73.5 ± 38.2	11
CHn1	47.7 ± 5.4	4

(a) Error bars indicate ± one standard deviation.   Source: CRWMS M&O (1997d)

Table 3-30. Average Shear Strength Parameters for Fractures from Each of the Four Thermal/Mechanical Units Sampled

Thermal/Mechanical Unit	Cohesion (C)[a] (MPa)	Coefficient of Friction[a] (tan ϑ)	Number of Tests
TCw	0.03 ± N/A	1.03 ± N/A	2
TSw1	1.63 ± 0.68	0.76 ± 0.08	16
TSw2	0.86 ± 0.81	0.87 ± 0.09	12
CHn1	1.69 ± 1.55	0.78 ± 0.25	4

(a) Error bars indicate ± one standard deviation.
N/A—Not applicable

Source: CRWMS M&O (1997d)

Table 3-31. Peak Shear Stress, Residual Shear Stress, and Applied Normal Stress for Fractures

Thermal/Mechanical Unit	Lithologic Unit	Specimen ID Number	σ (MPa)	τ_p (MPa)	τ_r (MPa)
TCw	Tpcpll	NRG-6-27.6-28.1-SNL	5.0	5.16	4.93
	Tpcplnc	NRG-6-102.9-103.4-SNL	15.0	15.46	10.20
TSw1	Tptrl	NRG-4-670.5-671.1-SNL	2.5	3.49	2.59
	Tptrn	NRG-6-380.9-381.3-SNL	2.5	3.68	2.95
	Tptpmn	SD-12-688.2-688.6-SNL	2.5	3.62	3.31
	Tptrn	NRG-4-537.8-538.2-SNL	5.0	5.48	5.21
	Tptrn	NRG-4-608.7-609.2-SNL	5.0	4.44	4.17
	Tptrn	NRG-6-297.4-297.7-SNL	5.0	5.16	5.01
	Tptrn	NRG-6-424.0-424.5-SNL	5.0	5.28	3.98
	Tptrn	NRG-7-367.2-367.8-SNL	5.0	6.47	4.2
	Tptrn	NRG-7-408.8-409.6-SNL	5.0	4.18	3.71
	Tptrn	NRG-6-296.0-296.4-SNL	10.0	11.89	6.59
	Tptrn	NRG-6-401.5-401.9-SNL	10.0	11.10	7.84
	Tptpul2	NRG-6-485.9-486.3-SNL	10.0	10.19	7.82
	Tptrn	NRG-7-434.7-435.3-SNL	10.0	6.85	6.02
	Tptrn	NRG-7-447.2-448.1-SNL	10.0	9.45	5.41
	Tptrn	NRG-7-317.3-317.8-SNL	15.0	11.50	10.70
	Tptrn	NRG-7-430.2-430.9-SNL	15.0	13.26	9.09
TSw2	Tptpmn	NRG-6-782.3-782.6-SNL	2.5	1.92	1.80
	Tptpln	SD-9-1255.9-1256.3-SNL	2.5	2.37	2.29
	Tptpln	SD-12-1072.5-1073.0-SNL	2.5	3.31	2.67
	Tptpll	SD-9-1132.2-1132.9-SNL	5.0	5.45	4.53
	Tptpll	SD-9-1171.1-1171.8-SNL	5.0	5.48	4.81
	Tptpln	SD-12-1072.5-1073.0-SNL	5.0	6.62	5.23
	Tptpln	SD-9-1254.7-1255.2-SNL	10.0	7.68	7.10
	Tptpln	SD-9-1254.7-1255.2-SNL	10.0	8.97	8.34
	Tptpmn	SD-12-778.1-780.0-SNL	10.0	12.00	10.38
	Tptpll	NRG-6-935.0-935.3-SNL	15.0	11.89	8.56
	Tptpll	SD-9-1141.2-1141.5-SNL	15.0	14.00	14.0
	Tptpll	SD-9-1144.2-1145.1-SNL	15.0	15.50	10.2
CHn1	Tac	SD-9-1480.7-1481.8-SNL	2.5	3.52	2.89
	Tac	SD-9-1480.7-1481.8-SNL	5.0	4.35	3.66
	Tac	NRG-7-1511.3-1512.0-SNL	5.0	7.03	4.81
	Tac	SD-9-1480.7-1481.8-SNL	10.0	9.45	5.46

(a) σ: normal stress; τ_p: peak shear stress; τ_r: residual shear stress

Source: CRWMS M&O 1997d

at the rock mass scale as inputs. Mechanical properties are known to be very different for strong, jointed, in situ rock masses than for small, intact samples tested in the laboratory. These differences are termed "scale effects" and are attributed to the influence of the size of the rock mass affected and to heterogeneities, such as jointing.

Rock Mass Classification

Rock mass quality data were collected in exploration boreholes and in the Exploratory Studies Facility. The rock mass quality system (Barton et al., 1974) and the rock mass rating system (Bieniawski, 1989) were employed as the basis of empirical design of excavation ground support and empirical correlation with rock mass properties. These two indices are rock classification methods that consider characteristics of the rock mass such as the degree of jointing, the interaction of joints to form blocks, joint surface frictional characteristics, rock strength, rock stress, and hydrologic conditions. Rock mass quality indices and the parameters used to determine the indices are not primary data, but they are derived from direct observations of rock mass characteristics.

The calculation of rock mass rating (RMR) requires six parameters that consider the strength of the rock, the rock quality designation (RQD), the joint spacing, the condition of joint surfaces, the groundwater environment, and a factor for the adjustment of joint orientation toward the excavation, as shown in the following equation:

$$RMR = C + I_{RQD} + JS + JC + JW + AJO , \qquad (3\text{-}6)$$

where RMR = a dimensionless number from 0 to 100, C = the strength parameter, I_{RQD} = the RQD parameter, JS = the joint spacing parameter, JC = the joint surface condition parameter, JW = the groundwater parameter, and AJO = the adjustment for joint orientation.

Parameter values are assigned based on classification guidelines presented by Bieniawski (1989) (Table 3-32). Adjustments for joint orientation can be made to the rock mass rating to account for effects of direction of mining approach. When application of the rock mass rating index is limited to estimation of rock mass mechanical properties in drift design methodology, adjustment for joint orientation is not applied. Borehole estimates of rock mass rating do not include adjustment for joint orientation because joint orientations cannot be determined from cores. The scan-line data include the adjustment for joint orientation factor, because joint orientations can in fact be evaluated with respect to the tunnel axis.

The rock mass quality index (Q), as defined by Barton et al. (1974), is calculated from six parameters:

$$Q = \left(\frac{RQD}{Jn} \right) * \left(\frac{Jr}{Ja} \right) * \left(\frac{Jw}{SRF} \right) . \qquad (3\text{-}7)$$

The first term (rock quality designation/Jn) describes the block size, the second term (Jr/Ja) describes interblock shear strength, and the third term (Jw/SRF) describes the effect of the active stress. Relative classes of rock quality based on the overall value of Q have been assigned by Barton et al. (1974) (Table 3-33).

Data and methodology used to estimate rock mass quality (Q) and rock mass rating (RMR) from the core data are described in Brechtel et al. (1995) and Kicker et al. (1996). In the Yucca Mountain methodology, rock mass quality was estimated for every 3 m (10 ft) interval of the core log. Some parameters used to calculate the rock mass quality or rock mass rating index could not be determined from the core, and values for each interval were therefore estimated by Monte Carlo simulations from distributions of the parameter, which were based on surface mapping and mapping of the North Ramp Starter Tunnel. The approach assumed that the value of each parameter was independent from the other parameters in each interval.

Rock mass quality data were also generated for 5 m (16.4 ft) intervals of the Exploratory Studies Facility, based on scan-line observations made on the excavation surface. "Scan line" refers to the determination of parameters along linear traces within the interval, as opposed to a detailed mapping of the features in the interval. The methodology used for the scan-line rock mass quality determinations is described in CRWMS M&O (1997d).

Both the Q and the RMR empirical rock classification systems were applied in the Exploratory Studies Facility, based on two separate sets of mapping data, including scan-line field mapping data and full-peripheral field mapping. Q and RMR were collected for each 5 m interval of tunnel. To smooth the spatial variability along adjacent 5 m intervals of the Exploratory Studies Facility, a nine-term moving average rock mass property value was determined for each interval, such that the value of a particular interval was averaged together with the values of the four preceding intervals and the four succeeding intervals.

RMR data (Table 3-34) typically show higher rock mass quality compared to Q data (Table 3-34). Scan-line and full-peripheral data are in very good agreement throughout the Exploratory Studies Facility, with the exception of the TSw1 thermal/mechanical unit in the North Ramp. A comparison of rock mass quality values in the TSw1 units shows that the greatest difference between the two data sources corresponds primarily to the Topopah Spring crystal-poor upper lithophysal zone (Tptpul). There is an observable structural difference along the tunnel alignment within the Tptpul in the North Ramp, such that the rock exposure is significantly smoother and less fractured above the spring line compared to below the spring line. The scan-line Q assessment was conducted above the spring line and therefore does not reflect the lower-quality rock below the spring line.

An empirical method for assessing the Q system parameter called the stress reduction factor (Kirsten, 1988) was applied to both scan-line and full-peripheral data sets in the Exploratory Studies Facility, resulting in the determination of a modified Q value. The method and analysis are described in detail by Kicker et al.

Table 3-32. Geomechanical Classification of Rock Masses Based on Rock Mass Rating

Rock Mass Class	Relative Rating	Range in RMR
I	very good	81 - 100
II	good	61 - 80
III	fair	41 - 60
IV	poor	21 - 40
V	very poor	<20

NOTE: Classification guidelines from Bieniawski (1979)

(CRWMS M&O, 1997f). The Q values calculated by the Kirsten approach, defined as $Q_{modified}$, are typically higher for both data sets and are generally in closer agreement to the RMR values.

This cumulative frequency of occurrence in the rock mass quality data set for each thermal/mechanical unit is summarized in Table 3-34, with frequencies of occurrence of 5%, 20%, 40%, 70%, and 90%. These frequencies correspond to the five rock mass quality categories as defined by Hardy and Bauer (1991) and serve as the basis for evaluating the potential range of rock mass conditions.

The most significant potential sources of bias are directional bias, bias related to the effect of scale (the smaller amount of structural data available from a borehole), and bias related to sample disturbance during borehole drilling. Comparison of rock mass quality projected from boreholes to the scan-line assessment data from the Exploratory Studies Facility excavations shows that the borehole data generally produce conservative estimates. The TCw is the only exception, where the impacts of normal faults on the rock are not well represented in the core data, as the predominantly vertical boreholes do not provide a good sampling of fault impact on the brittle tuff rocks.

Borehole rock quality designation (RQD) is much lower than the rock quality designation as assessed at the tunnel scale. This is related to the character of the core recovered from drilling. The tuffs are extensively fractured at the core scale, which causes a large amount of rubble, lost core, and high fracture frequency. Attempts to filter drilling-induced fractures from the rock quality designation calculations (enhanced–rock quality designation) increases the values by factors of 1.5–2.2 for different thermal/mechanical units. However, rock quality designation values from the tunnel scale are still greater by a factor of 1.5–3.1. The RQD in the tunnel assessments is based on frequency of observable fractures, which excludes many fractures that impact the core. In addition, although the extent of rubble or highly fractured intervals is included in the tunnel assessment of RQD, those types of features are much less common than suggested by the core. Joint frequencies are higher in core data because core data include smaller-scale fractures that are not counted at the tunnel scale. This resulted in small values of the joint spacing parameter in the core rock mass rating. Joint set numbers in the core are based on fractures observed in the

Table 3-33. Relative Rock Quality for Ranges of Q

Relative Rating	Q
Exceptionally Good	400 - 1000
Extremely Good	100 - 400
Very Good	40 - 100
Good	10 - 40
Fair	4 - 10
Poor	1 - 4
Very Poor	0.1 - 1
Extremely Poor	0.01 - 0.1
Exceptionally Poor	0.001 - 0.01

Source: Barton, N.R. et al. (1974)

TCw, which has both more joint sets and higher frequencies of jointing than other units. Jn values used for the PTn and TSw1 are higher than revealed by the excavation.

The lowest rock mass quality is observed in the TCw thermal/mechanical unit (Table 3-34), which also has the greatest variability. Rock mass quality is lowest in the most densely welded lithostratigraphic units, Tpcpul and Tpcpmn, in the Tiva Canyon Tuff (CRWMS M&O, 1996b). It is higher in the less densely welded upper and lower portions. This correlates with analysis of the fracture mapping for TCw, which consistently indicates more joint sets and higher joint frequency. In addition, the North Ramp penetrates the TCw in a zone of normal faulting, which contributes to the broken character of the TCw rocks (CRWMS M&O, 1996b).

The PTn unit shows consistently high rock mass quality ratings (Table 3-34). Typically, only one set of joints is evident in the PTn and has very limited impact on the excavation. Rock strength is low in this unit, with some intervals that are nonlithified. Shear failures are observed on the sides of the tunnels in some of the weaker PTn materials. However, they are localized and have not affected the long-term character of the excavation.

Table 3-34. Summary of Rock Mass Quality Values

Thermo/ Mechanical Unit	Rock Mass Quality Category	Cumulative Frequency	Rock Mass Rating (RMR)			Rock Mass Quality, Q			Q_modified (Kirsten Approach)	
			NRG Borehole[1]	Scanline	Full-Peripheral	ESF Design[2] (NRG Boreholes)	Scanline[3]	Full-Peripheral[3]	Scanline[3]	Full-Peripheral[3]
TCw	1	5%	51.00	47.00	43.00	0.38	0.21	0.33	0.99	0.46
	2	20%	56.00	53.80	54.00	0.68	0.44	0.81	2.35	2.27
	3	40%	59.00	58.80	60.00	2.08	1.01	1.80	4.96	4.39
	4	70%	67.00	66.60	66.00	5.66	5.22	4.15	18.12	9.60
	5	90%	72.00	74.10	74.00	9.14	38.44	14.50	67.78	22.24
UO	1	5%	50.40	48.90	51.00	--	5.03	2.00	7.79	3.90
	2	20%	52.90	62.60	59.00	--	18.67	16.00	8.62	7.19
	3	40%	55.88	66.70	64.00	--	30.00	24.00	12.55	9.69
	4	70%	59.42	72.60	86.00	--	38.00	33.00	15.84	14.58
	5	90%	60.00	87.00	86.00	--	625.00	42.00	244.64	17.36
PTn	1	5%	45.00	53.70	56.00	0.15	1.54	1.28	0.52	0.32
	2	20%	52.00	61.20	61.00	0.28	5.56	4.20	1.11	0.70
	3	40%	55.00	71.10	70.00	0.66	7.80	9.00	2.30	0.98
	4	70%	65.00	74.00	77.00	1.62	13.11	13.00	10.59	1.43
	5	90%	70.00	87.00	80.00	3.74	616.88	25.00	41.97	3.50
TSw1	1	5%	49.00	48.10	41.00	0.24	0.38	0.48	1.35	0.77
	2	20%	53.00	55.10	48.00	0.87	2.07	1.60	5.78	2.59
	3	40%	57.00	61.40	54.00	1.73	5.13	3.70	13.55	4.79
	4	70%	62.00	74.10	63.00	5.09	31.90	10.00	67.00	9.66
	5	90%	70.00	92.00	70.00	12.00	231.50	22.00	372.90	18.34
TSw2	1	5%	51.00	53.00	44.00	0.30	0.55	0.47	3.26	0.82
	2	20%	56.00	58.10	52.00	0.65	1.18	1.03	6.22	1.90
	3	40%	58.00	62.10	56.00	1.91	2.78	1.70	11.46	3.43
	4	70%	63.00	66.10	62.00	3.75	6.18	4.10	25.47	6.99
	5	90%	67.00	72.10	68.00	8.44	16.28	9.30	56.14	15.07

Source: CRWMS M&O 1997f

NOTES: 1. From SNL (1995b), "Data Transmittal for Rock Mass Mechanical Properties Estimates from NRG Drilling Program," TDIFs 304414 and 304415; DTNs SNF29041993002.062 and SNF29041993002.063.

2. From CRWMS M&O (1995b), "ESF Ground Support Design Analysis," BABEE0000-01717-0200-00002 Rev.00.

3. Values of Q are extracted from rank-ordered intervals whose cumulative frequency is closest to the corresponding rock mass quality category cumulative frequency.

Rock mass quality is higher in the TSw1 unit than in the corresponding portions of the TCw (CRWMS M&O, 1996c). Jointing is less well developed. Poor rock mass quality, anticipated in the upper lithophysal zone (Tptpul), is not observed. Jointing is not well developed and is generally limited to one set. The heterogeneities in the Tptpul caused by large lithophysae and relatively small cracking of the rock has little effect on the rock mass at the excavation scale. Where the middle nonlithophysal zone (Tptpmn) in the TSw2 is exposed in excavations of the Main Drift, rock mass quality is relatively high.

Rock Mass Thermal Properties

Correlations have been developed or proposed for thermal/mechanical properties at the rock mass scale (Nimick and Connolly, 1991). Thermal conductivity at the intact scale has been shown to be a function of porosity, saturation, and temperature. Differences at the rock mass scale are projected to be related to the additional fracture porosity, which should be a small effect. Similarly, the heat capacity of intact rock is expected to be an adequate predictor of heat capacity at the rock mass scale.

Preliminary thermal/mechanical analyses for design have been performed in an attempt to project laboratory thermal expansion data to the rock mass scale, as described in Jung et al. (1993). The preliminary thermal/mechanical analyses indicate a maximum upward displacement of almost 30 cm at the surface, 300 yr after waste emplacement. Most of this displacement would originate in the TSw2 unit. The rock in the immediate vicinity of the repository is predicted to be in compression, but the tensile stress nearer the surface (TCw thermal/mechanical unit) is predicted to be relatively high (~5 MPa). This behavior

could potentially result in the opening of preferential pathways for water infiltration or gas migration (CRWMS M&O, 1997d).

Data collected from the Single Heater Test through May 1997 (CRWMS M&O, 1997g) indicate that the temperature distribution around the heater is radially symmetric and that conduction thus appears to be the primary mode of heat transfer through the rock mass. However, some anomalous temperature gauge readings may indicate convective heat transport in fractures. The available thermal data also indicate the formation of a dry-out zone extending radially outward roughly 1 m from the heater to approximately the 100 °C isotherm.

The thermal expansion coefficient of the rock mass can be determined from selected multipurpose borehole extensometer displacements and temperatures. The calculated rock mass thermal expansion coefficient ranges between ~4 and $6 \times 10^{-6}/°C$. Rock mass thermal expansion can be calculated from the in situ data, including temperature change for a given axial length from ambient, gauge length, and measured thermal displacement over the gauge length (Tables 3-35 and 3-36).

Rock Mass Mechanical Properties

Rock Mass Strength

Rock mass mechanical properties have been estimated using the approach proposed by Hardy and Bauer (1991), using laboratory test data and rock mass quality rating to estimate mechanical properties at the rock mass scale for use in equivalent continuum analyses. The estimated properties are listed in Table 3-37 for each thermal/mechanical unit and rock mass rating values at 40% cumulative frequency of occurrence. Ranges of the rock mass properties are estimated based on rock mass rating from

Table 3-35. Calculated Rock Mass Thermal Expansion Coefficient from SHT Data through Day 90

MPBX Number	Average ∝ $(10^{-6}/°C)$	Average Temperature (°C)	Maximum Gage Length (m)
TMA-MPBX-1	5.02	118.64	2.0
TMA-MPBX-3	5.27	63.81	3.0

∝ = Coefficient of thermal expansion

Table 3-36. Thermal Expansion Coefficients from SHT Data for Longest Available Gage Lengths, Near, Heating Cycle

MPBX Number	Anchor Numbers	Average ∝ $(10^{-6}/°C)$	Average Temperature (°C)	Gage Length (m)
TMA-MPBX-1	1 to 4	5.88	160.3	2.84
TMA-MPBX-3	2 to 6	4.14	70.07	4

∝ = Coefficient of thermal expansion

Table 3-37.   Estimated Rock Mass Mechanical Properties for Rock Mass Rating at 40 Percent Cumulative Frequency of Occurrence for Each Thermal/Mechanical Unit

Rock Mass Mechanical Property	T/M Unit				
	UO	TCw	PTn	TSw1	TSw2
Strength = $A+B\sigma_3{}^C$					
Uniaxial Compressive Strength -A (MPa)	0.88	14.4	1.51	6.4	25.4
B	12.018	14.337	9.585	9.271	23.687
C	0.729	0.709	0.719	0.712	0.571
Mohr-Coulomb Failure					
Cohesion (MPa)	1.1	5.5	1.3	3.3	8.9
Angle of Internal Friction (degrees)	52	55	48	47	59
Dilation Angle (degrees)	26	27	24	24	30
Deformation Modulus (GPa)	4.3	25.7	2.5	18.9	31.1
Poisson's Ratio	0.14	0.21	0.2	0.26	0.21

Source: CRWMS M&O (1997d)

scan-line data and the average of the appropriate intact rock property. Two sets of empirical rock mass strength criteria, Yudhbir and Prinzl (1983) and Hoek and Brown (1988), were adopted for the drift design methodology, and an average of the two predicted strengths was used to develop a power-law relationship of rock mass strength versus confining pressure. Information required for obtaining rock mass strength includes rock mass quality indices, intact rock uniaxial compressive strengths, and the triaxial compressive strength data. Design parameters for rock mass elastic modulus (Serafim and Pereira, 1983), Poisson's ratios, and Mohr-Coulomb strength were also developed for each thermal/mechanical unit.

The equation proposed by Yudhbir et al. (1983) for calculation of rock mass strength is:

$$\sigma_1 = A\sigma_c + B\sigma_c \left(\frac{\sigma_3}{\sigma_c}\right)^\alpha, \qquad (3\text{-}8)$$

where σ_c = intact rock uniaxial compressive strength, σ_1 = strength of the rock mass, σ_3 = confining stress, A = a dimensionless parameter dependent on the rock mass rating, and α, B = rock material constants dependent on rock type.

The value of A for the rock mass is obtained from the design rock mass rating (RMR_D) by the following equation according to Yudhbir et al. (1983):

$$A = e^{0.0765(RMR_D)-7.65}. \qquad (3\text{-}9)$$

The material constants B and α are related to the rock type and are determined by a curve fitting of the confined compressive

strength test results (Table 3-38). For the TCw thermal mechanical unit, NRG core triaxial test data were used to determine B and α using the method outlined in Hardy and Bauer (1991) and Lin et al. (1993). These data were originally published in Brechtel et al. (1995). For the undifferentiated overburden and PTn thermal/mechanical units, NRG uniaxial compression and Brazilian tensile strength tests were used to determine B and α with modifications of the method suggested by Hardy and Bauer (1991) and Lin et al. (1993). These data were also originally published in Brechtel et al. (1995). For the TSw1 and TSw2 thermal/mechanical units, triaxial test data from samples were used to determine B and α using the method outlined in Hardy and Bauer (1991) and Lin et al. (1993). Only five data points for each unit were available to calculate these constants.

The Hoek and Brown (1988) rock mass strength criterion is shown here:

$$\sigma_1 = \sigma_3 + \sqrt{m\sigma_c\sigma_3 + s\sigma_c^2}, \qquad (3\text{-}10)$$

where m = a constant that depends on the properties of the rock, and $m = m_i e^{(RMR-100)/28}$, s = a constant that depends on the extent to which the rock is fractured, and $s = e^{(RMR-100)/9}$.

The parameter m_i is the constant for intact rock determined by curve fitting of the confined compressive strength test data (Table 3-38). The design rock mass strengths for each rock mass quality category were calculated by averaging the strengths determined from both Yudhbir et al. (1983) and Hoek and Brown (1988) criteria, following the procedure of Hardy and Bauer (1991).

Table 3-38. Intact Rock Constants for Rock Mass Strength Criteria

Thermal/Mechanical Unit		Yudhbir Criterion		Hoek & Brown Criterion	Intact Rock Compressive Strength (MPa)
		B	α	m_i	σ_c
UO (Tuff "X")		8.10	1.00	125.64	6.8
TCw		2.50	0.64	18.50	125.1
PTn	Tiva Canyon, Yucca Mountain	4.56	1.00	17.64	8.0
	Bedded Tuffs, Pah Canyon	6.10	1.00	150.97	
TSw1		2.62	0.663	10.60	56.9
TSw2		1.63	43	19.68	178.5

Source: CRWMS M&O (1997d)

A power-law relationship of the form

$$\sigma_1 = A + B\sigma_3^C \qquad (3\text{-}11)$$

was employed to describe the nonlinear design rock mass strength. The parameters A, B, and C were determined by curve fitting the strength envelopes using a least square method, and are included in Table 3-37 for 40% cumulative frequency for each thermal/mechanical unit.

The Mohr-Coulomb strength parameters, including cohesion, angle of internal friction, and the dilation angle, are commonly used to describe rock mass strength in numerical analysis. The strength parameters were developed from the least square curve fits of strength data pairs (σ_1, σ_3) produced using the power-law criterion described above and summarized for 40% frequency in Table 3-37. The linear relation for strength (σ_1) and confining pressure (σ_3) is defined in the form of the following equation:

$$\sigma_1 = \sigma_c + N\sigma_3, \qquad (3\text{-}12)$$

where σ_c = uniaxial compressive strength, and N = confinement factor.

The parameters σ_c and N were then used to create a Mohr-Coulomb failure criterion relating the shear and normal stress on the plane of failure to cohesion and angle of internal friction by the following equation:

$$\tau = C_0 + \sigma_n \tan\phi, \qquad (3\text{-}13)$$

where C_0 = cohesion, and $C_0 = \sigma_c/\sqrt{N}$, ϕ = angle of internal friction, and $\phi = 2\left(\tan^{-1}\sqrt{N} - 45°\right)$.

The least square best fit was performed over the range of confining pressures from 0 to 3 MPa, which is representative of the projected range in minimum principal stresses near the boundary of the excavations and includes the resulting Mohr-Coulomb strength parameters for 40% cumulative frequency of occurrence (Table 3-37).

The nonassociated flow rule, suggested by Michelis and Brown (1986), which uses a dilation angle equal to half the internal friction angle, was considered suitable for the tuff (Hardy and Bauer, 1991) (Table 3-37).

Two additional empirical methods were used for assessing rock mass strength properties: the Geological Strength Index (Hoek and Brown, 1988) and the Rock Mass Index (Palmstrom, 1996). Properties estimated included rock mass elastic modulus, cohesion, friction angle, unconfined compressive strength, tensile strength, joint cohesion, and friction angle.

Rock Mass Elastic Moduli

Serafim and Pereira (1983) developed a correlation between the rock mass rating and rock mass elastic modulus that was recommended for use by Hardy and Bauer (1991). The correlation is shown in the following equation:

$$E = 10^{\frac{(RMR-10)}{40}}, \qquad (3\text{-}14)$$

where E is in GPa.

Because the equation does not incorporate the intact rock elastic modulus, the predicted rock mass elastic modulus can exceed the intact rock elastic modulus at high rock mass rating values. An upper limit of the rock mass modulus was, therefore, set equal to the intact rock modulus (Table 3-38). Calculated rock mass moduli based on design rock mass rating values are shown for 40% cumulative frequency of occurrence in Table 3-37.

Rock mass moduli were determined using Geological Strength Index and Rock Mass Index indices and resulted in modulus values that significantly exceeded the mean intact value for the undifferentiated overburden, PTn, and TSw1 thermal/mechanical units based on field mapping data (CRWMS M&O, 1997f). The Rock Mass Index empirical methodology applied in this analysis resulted in a significantly smaller range of E values that is more consistent with the intact value for these units.

Empirical relationships to estimate Poisson's ratio from rock mass quality are not available. The mean values for intact rock Poisson's ratios from the laboratory tests for each thermal/mechanical unit were adopted as the rock mass Poisson's ratios (Table 3-37). No adjustments for rock mass category are recommended.

Rock mass elastic moduli were also determined in situ as part of the Single Heater Test, using the NX borehole jack (Goodman Jack). Results, procedures, and analysis are described in SNL (1997b) and in CRWMS M&O (1997h). This nonpermanent borehole instrument was periodically inserted into a single borehole drilled roughly horizontal and perpendicular to the Single Heater Test heater and pressurized at various distances along the borehole. Jack pressure and loading platen displacements were monitored, and rock mass modulus was determined from the pressure/displacement curve.

The NX borehole jack consisted of two hydraulically activated steel loading platens ~20.3 cm long, which applied a unidirectional load to a nominal 7.62-cm-diameter borehole wall. The maximum jack pressure was 69 MPa, and the maximum platen displacement was 0.63 cm. Jack pressure was applied using an Enerpak hand pump. The jack was inserted into the borehole, and platens were slowly expanded until the pressure just began to rise. The resulting linear variable displacement transducer readings represent initial borehole diameter and were used for calculations of borehole wall displacement under pressure. The jack pressure was increased in increments to the desired maximum pressure and then decreased in similar increments. Typically, the jack was pressurized in 3.44 MPa (500 psi) increments to 55.2 MPa (8000 psi), then back to zero, with linear variable displacement transducer readings recorded during both loading and unloading.

Rock modulus values ranged from ~3 to 23 GPa (Table 3-39). The data show thermally induced stiffening of the rock mass in the region near the heater, with rock mass moduli in this region increasing from 8 GPa in November 1996 to 23 GPa in March 1997. This modulus increase is likely because of the closing of fractures by rock matrix thermal expansion (CRWMS M&O, 1997h).

Ambient in situ rock mass moduli calculated from Single Heater Test borehole jacking are lower than laboratory values determined for intact specimens, which are in turn lower than values determined from Q/RMR rock mass quality estimates. This is consistent with previous in situ experiments conducted in welded tuff in G-tunnel, which indicated that the modulus values for in situ tests were about half the intact laboratory-determined value of ~23–35 GPa (Zimmerman and Finley, 1987).

IN SITU STRESS CONDITIONS

The direction of the maximum principal in situ stress at the repository horizon is vertical, because of the lithostatic load (Table 3-40). At the repository level, the vertical stress has been assumed to be 7.0 MPa on the average (Stock et al., 1984;

CRWMS M&O, 1995). Horizontal stresses are expected to be lower and to range from 3.5 MPa to 4.2 MPa, although the range may be as wide as 2.1 MPa to 7.0 MPa. These in situ stress values were generally confirmed by a stress profile calculated for the Exploratory Studies Facility test area (YMP, 1995), which shows a vertical stress of 6.0 MPa at a depth of 300 m. Horizontal stress for the same depth ranges from 2.1 to 4.2 MPa.

Horizontal in situ stresses at the repository site are expected to be generally low. Consequently, failure modes around underground openings during construction are expected to be primarily controlled by geologic structures. Minimum and maximum horizontal/vertical stress ratios are close, indicating a weak horizontal stress anisotropy. Lateral stresses and their effects would thus be expected to be similar for all drift orientations (CRWMS M&O, 1997d).

A series of five successful hydraulic fracturing tests was conducted in the Drift Scale Test block, but only one test yielded what were considered reliable results. Based on these test results, the principal horizontal stresses around this borehole are estimated to be (SNL, 1997a):

$$\sigma_h = 1.7\ (\pm 0.1)\ \text{MPa},$$
$$\text{acting in the N75°W } (\pm 14°) \text{ direction} \qquad (3\text{-}15)$$

and

$$\sigma_H = 2.9\ (\pm 0.4)\ \text{MPa},$$
$$\text{acting in the N15°E } (\pm 14°) \text{ direction,} \qquad (3\text{-}16)$$

where σ_h = least horizontal principal stress, and σ_H = largest horizontal principal stress.

Because vertical stress was not measured in these tests, it was approximated, as the weight of the overburden at the depth of the tests from the surface, using (SNL, 1997a):

$$\sigma_v = 4.7\ \text{MPa}, \qquad (3\text{-}17)$$

where σ_v = vertical stress.

Although the measured horizontal stresses are only moderately differential, both are smaller than the vertical stress. This measured stress regime, one of low horizontal magnitudes, is in accord with the dominant local normal faults. The N-NE maximum horizontal stress direction is subparallel to the average strike of these faults and is supported by previous measurements in the Yucca Mountain area (Zoback and Healy, 1984).

EXCAVATION CHARACTERISTICS

Specific safety and health concerns related to the mineralogy of tuffs at Yucca Mountain include respiratory effects of erionite and silica minerals (including quartz and cristobalite) during daily underground activities. These minerals occur in varying proportions in the different lithologies and geochemical environments at Yucca Mountain as reported in Vaniman et al. (1996).

Table 3-39. **Calculate Rock Mass Modulus in Borehole ESF-TMA-BJ-1 Using the Borehole Jack for the Single Heater Test**

| Date | Distance from Collar | | | | |
| | 2.0 m | 3.0 | 4.0 m | 4.51 m | 6.2 m |
	Rock Mass Modulus GPa (Temp °C)				
8/26/96	6.9 (25)	3.71 (25)	No test	No test	No test
10/10/96	10.3 (27.5)	10.3 (27.7)	8.3 (30.2)	6.0 (34)	No test
11/26/96	Results discarded (31.1)	10.2 (35.9)	5.71 (46.4)	5.01 (55.4)	8.4 (141.8)
3/18/97	Results discarded (35)	6.3 (41)	10.3 (52)	5.7 (58.7)	22.8 (143.1)

Source: CRWMS M&O (1997d)

NOTE: Calculated moduli are based on field data in which the difference between the two borehole jack LVDT readings slightly exceeded the limits set in ASTM D4971-89. The fractured nature of the rock made setting the jack difficult. Discarded results were for data that far exceeded ASTM D4971-89 limits.

Table 3-40. Summary of In Situ Stresses at the Repository Horizon

Parameter	Average Value	Range of Values
Vertical Stress	7.0 MPa	5.0 - 10.0 MPa
Minimum Horizontal/Vertical Stress Ratio	0.5	0.3 - 0.8
Maximum Horizontal/Vertical Stress Ratio	0.6	0.3 - 1.0
Bearing of Minimum Horizontal Stress	N57°W	N50°W - N65°W
Bearing of Maximum Horizontal Stress	N32°E	N25°E - N40°E

Source: *Advanced Conceptual Design Report* (CRWMS M&O, 1996f)

Hazards include erionite, a carcinogen, and crystalline silica, which can produce respiratory ailments upon becoming airborne and during tunneling operations. Control measures regarding these mineralogic respiratory hazards are performed according to Harris (M.W. Harris, February 1995, written communication to W.R. Dixon) for erionite and McManus (1996) for silica. These measures include monitoring, engineering control, and proper personal protective equipment when tunneling activities are occurring in areas where these minerals are of concern.

Excavation Methods

The North Ramp Starter Tunnel, Upper Tiva Canyon Alcove, Alcove 2, and the Thermal Test Facility (Thermal Testing Facility Alcove) were excavated by drilling and blasting. Controlled blasting procedures were implemented to minimize blast-induced damage to the excavation perimeter. Blast damage can result in loosening of the surrounding rock mass, which increases ground support requirements and long-term maintenance.

The Nuclear Regulatory Commission calls for an "excavation method that will limit the potential for creating a preferential pathway for groundwater or radioactive waste migration to the accessible environment." The Exploratory Studies Facility Design Requirements (YMP, 1995) estimated that the rock mass altered by the excavation process would be within 1.5 m of the excavated surface.

Additional criteria for blast vibration limits, based on measurements of peak particle velocity, were developed for the excavation perimeter and nearby structural elements (Table 3-41). These vibration criteria are specified without reference to the dominant frequency. Application of peak particle velocity–distance relationships is required to obtain estimated peak particle velocity within 1 m of the blast. Far-field monitoring utilized seismic equipment with peak frequency ranges of 250 Hz.

Excavation of the North Ramp Starter Tunnel and Upper Tiva Canyon Alcove produced substantial overbreak on existing structural features, even with the use of perimeter blasting procedures. Monitoring of the blasts indicated that reduced quantities of explosives detonated per delay and long-period delays were successful in controlling peak particle velocities. Observations suggested that measurable blast damage was limited to within 1 m of the excavation perimeter.

Table 3-41. Blast Vibration Limits Specified for Construction of Alcove 2

Structure	Peak Particle Velocity	
	(mm/s)	(in/s)
Opening Perimeter at 1 m Distance	700	28
Steel Sets	1250	50
Cast-in-Place Concrete (Age: 0-24 hours)	No Blasting	No Blasting
Cast-in-Place Concrete (Age: 24-60 hours)	10	0.4
Cast-in-Place Concrete (Age: >60 hours)	102	4

Source: CRWMS M&O (1996c)

During construction of Bow Ridge Fault Alcove and a portion of the Thermal Test Facility (Thermal Testing Facility Alcove), construction monitoring consisted of monitoring ground motion (peak particle velocity), making observations of damage in the rock surrounding the excavation, and making visual observations of the number of borehole half-casts remaining on the excavation perimeter. Half-casts of perimeter trim holes were mapped on the east and west walls of Upper Tiva Canyon Alcove. Half-casts were rare on the east wall. They occurred more frequently on the west wall, but were irregular in occurrence. The half-cast distribution may be attributed to drilling and blasting practices or to structural geologic controls, such as joint spacing and orientation (CRWMS M&O, 1997a).

Geophones were attached to steel set no. 90, adjacent to the entrance of Upper Tiva Canyon Alcove, and recorded a peak particle velocity of 330 mm/s (13 in/s) in the direction of the North Ramp axis during blasts no. 5 and 6. These geophones measured blast vibrations with frequencies below 250 Hz and peak particle velocities well below the 1250 mm/s criterion (Table 3-41).

Visual observations of blasting damage at Upper Tiva Canyon Alcove were made by video borescope records of inspection boreholes at distances of 0.9, 1.5, and 2.5 m from the excavation perimeter. On the basis of observed debris in the holes, rock damage in the holes, and overbreak of the excavation, the blasting effects extended up to 1 m from the excavation perimeter. No damage was evident at 1.5 m from the perimeter.

A similar monitoring program, performed during construction of part of the Thermal Test Facility (Alcove 5), successfully produced both near-field and far-field blasting vibration data. An attenuation relationship was developed by combining the two data groups and was used to project the peak particle velocity at a distance of 1 m from the excavation perimeter for comparison to the criteria in Table 3-41. The calculation indicated that rock 1 m from the excavation perimeter would be most strongly impacted by detonation of the trim holes because they are closest and contain the highest quantity of explosive because of the number of holes. Application of the attenuation relationship predicted peak particle velocities between 703 and 720 mm/s, which would exceed the criterion of 700 mm/s at 1 m by a maximum of 3%.

Data scatter in the peak particle velocities used to develop the attenuation relationship is very large, and it is therefore highly likely that the criterion was not exceeded. Furthermore, the majority of the data were developed from blast holes where the explosive was tamped and fully coupled to the hole perimeter. Trim holes were loaded with decoupled charges (smaller diameter than the hole) and were not tamped. Peak particle velocities predicted for all other holes were well below the 700 mm/s criterion.

Blast monitoring results at the Thermal Test Facility suggest that damage to the rock was limited to within 1.5 m of the excavation boundary. Peak particle velocities at 1 m from the excavation boundary appeared to satisfy the blast vibration limits.

Excavation Characteristics

Ground support installed in the Exploratory Studies Facility includes rock bolts, lattice girders, steel fiber reinforced shotcrete, and steel sets (CRWMS M&O, 1997d). Wire mesh and channel straps were used to control loose materials between rock bolts. Monitoring of rock bolts was accomplished using rock bolt load cells and instrumented rock bolts. Convergence pins were attached to the lattice girders in the first 10 m of the Exploratory Studies Facility to monitor the displacement of these components of the ground support system. Vibrating wire strain gauges and convergence pins were attached to steel sets throughout the Exploratory Studies Facility to monitor the changes in rock loading. Convergence pin arrays and borehole extensometers were installed in rock supported by Swellex bolts (CRWMS M&O, 1997d).

Rock Bolt Load

Rock bolt load cells were installed along the North Ramp Starter Tunnel and on the highwall at the North Ramp portal. Installation procedures are detailed in SNL (1995). All bolts had some load bleed-off and have settled into generally stable trends, in which bolt loads are relatively constant. No load increases that would indicate rock loosening were observed from the time of installation to June 1996. Current bolt loads range from 0.1% to 16.0% of the bolt yield strength (CRWMS M&O, 1997d).

Table 3-42. Rock Bolt Load Cells, Load Versus Time

	TMA RLBC Gage	| Days After Startup — Average Load (lb.)									
		0	14	28	42	56	70	84	96	112	126
Hot Side	RB-LC-1-AVG	22662	22262.8	22158	21732.3	21537.1	21444.1	21407.5	21380.8	21340.3	21308.5
	RB-LC-2-AVG	14859.4	14739.7	14708.6	14680.1	1463.7	14597	14559.8	14522.5	14496.5	14449.6
	RB-LC-3-AVG	22428	22402.2	22378.7	22348.4	22317.5	22281	22262.3	22243.2	22231	22224.1
	RB-LC-4-AVG	16663.9	16602.8	16580.3	16558.8	16522.1	16496.6	16647.4	16446.3	16424.2	16407.5
Ambient Side	RB-LC-5-AVG	25971.9	25928.5	25887	25856.6	25829.3	25802.6	25783.4	25765.5	25748.7	25738.1
	RB-LC-6-AVG	14642.7	14633.2	14632.7	14627.3	14619.4	14609.5	14601.2	14595.9	14589.2	14573.7
	RB-LC-7-AVG	4932.6	4921.1	4919.7	4911.8	4907.3	4893.6	4890.9	4883.8	4877.5	4873
	RB-LC-8-AVG	16862.8	16818.5	16783.6	16738.7	16738.7	16605	16592.7	16575.4	16566	16561.5

	TMA RLBC Gage	| Days After Startup — Average Load (lb.)									
		140	154	168	182	196	210	224	238	252	266
Hot Side	RB-LC-1-AVG	21279.7	21254.3	21206.3	21176.9	21161.2	21145.9	21127.1	21112.2	21100.9	21102.1
	RB-LC-2-AVG	14422.7	14405.6	14389.9	14378.6	14369.9	14365.5	14353.4	14349	14342	14341.1
	RB-LC-3-AVG	22214.2	22206.8	22201.1	22194.3	22189.6	22183.4	22176.4	22171.7	22165.3	22158.4
	RB-LC-4-AVG	16394.3	16377.4	16361.5	16350.8	16340.4	16331	16320.2	16316.8	16312.1	16310.9
Ambient Side	RB-LC-5-AVG	25728.1	25722.2	25714.1	25705.1	25698.3	25692.7	25683.1	25676	25665.6	25652
	RB-LC-6-AVG	14567.1	14563.5	14562.3	14557.4	14553.9	14551.2	14549.3	14543.8	14543.4	14538.9
	RB-LC-7-AVG	4866.9	4866.7	4867.2	4866.6	4868.2	4865.2	4863.2	4863.9	4864.1	4867.1
	RB-LC-8-AVG	16552.8	16544.8	16538	16533.	16528.6	16522.3	16516.4	16514	16503.2	16501.5

Instrumented rock bolts were installed in Upper Tiva Canyon Alcove, as described in SNL (1995). Stable bolt loads are similar to those observed in rock bolt load cells. Bolt loads appeared to remain well below the bolt yield strength (SNL, 1995).

Eight rock bolt load cells were also installed as part of the Single Heater Test to evaluate the effects of elevated temperature on bolt performance. Complete data and analysis are presented in SNL (1997b) and CRWMS M&O (1997h). Four of the rock bolts were installed on the heated side of the thermomechanical alcove below the level of the heater and four additional bolts were installed on the opposite, cool side of the alcove. The load cells each contained three strain gauges, and the total load acting on the cell was calculated by averaging the measurements from all three.

The Single Heater Test rock bolt load cell data are presented in Table 3-42 as load and time from the start of heating, or day zero. There is a general decrease in load measured in all the load cells, although the decreases are all <7% of the initial load (CRWMS M&O, 1997h). The average percent decrease is 1.37% for rock bolts on the ambient (cool) side and 3.45% for rock bolts on the heated side.

Portal Girder Deformation

Deformation of the portal lattice girders embedded in shotcrete was tracked using convergence data collected by a tape extensometer. The deformations have remained fairly constant after the initial settling (SNL, 1995), and the monitoring data from June 1995 to June 1996 show a continuing trend of no closure (CRWMS M&O, 1997a).

Steel Set Deformation

Vibrating wire strain gauges were installed on 33 steel sets from January 1995 through June 1996. The strain gauges were attached to the steel sets both prior to and after installation in the tunnel. When the gauges were attached prior to installations, stress changes in the web of the steel sets due to jacking loads were monitored during the steel set installation process.

Strain changes that occurred during jacking installation indicated a generally similar pattern of tensile and compressive stress change in the steel. Changes in the crown of the steel set were the most uniform. The measured strain changes during jacking indicated stress changes between 10 and 180 MPa (CRWMS M&O, 1997a).

Strain magnitudes remained generally constant after installation was complete and suggest that loading of rock around the steel sets monitored was not occurring. Most of the steel sets were installed in the TCw and in the undifferentiated overburden units. Similar long-term trends indicating no increase of steel set load were observed for steel sets in the PTn, TSw1, and TSw2 units (CRWMS M&O, 1997d).

REFERENCES CITED

Arnold, B.W., Altman, S.J., Robey, T.H., Barnard, R.W., and Brown, T.J., 1995, Unsaturated-zone fast-path flow calculations for Yucca Mountain groundwater travel time analyses (GWTT-94), SAND95-0857: Albuquerque, New Mexico, Sandia National Laboratories.

American Society for Testing and Materials Standards (ASTM) C135–86, 1992, Test method for true specific gravity of materials by water immersion: Philadelphia, Pennsylvania, American Society for Testing and Materials.

ASTM D854–92, 1993, Standard test method for specific gravity of soils: Philadelphia, Pennsylvania, American Society for Testing and Materials.

ASTM D2664–86, 1993, Standard test method for triaxial compressive strength of undrained rock core specimens without pore pressure measurements: Philadelphia, Pennsylvania, American Society for Testing and Materials.

ASTM D3148–96, 1993, Standard test method for elastic moduli of intact rock core specimens in uniaxial compression: Philadelphia, Pennsylvania, American Society for Testing and Materials.

ASTM D3967–95, 1993, Standard test method for splitting tensile strength of intact rock core specimens: Philadelphia, Pennsylvania, American Society for Testing and Materials.

ASTM D4406–84, 1993, Standard test method for creep of cylindrical rock core specimens in triaxial compression: Philadelphia, Pennsylvania, American Society for Testing and Materials.

ASTM D4535–85, 1993, Standard test methods for measurement of thermal expansion of rock using a dilatometer: Philadelphia, Pennsylvania, American Society for Testing and Materials.

ASTM D4611–86, 1993, Standard practice for specific heat of rock and soil: Philadelphia, Pennsylvania, American Society for Testing and Materials.

ASTM D4971–89, 1991, Standard test method for determining the in situ modulus of deformation of rock using the diametrically loaded 76-mm (3-in.) borehole jack: Philadelphia, Pennsylvania, American Society for Testing and Materials.

ASTM E228–85, 1989, Standard test method for linear thermal expansion of solid materials with a vitreous silica dilatometer: Philadelphia, Pennsylvania, American Society for Testing and Materials.

ASTM F433-77, 1993, Standard practice for evaluating thermal conductivity of gasket materials: Philadelphia, Pennsylvania, American Society for Testing and Materials.

Barton, N.R., Lien, R., and Lunde, J., 1974, Engineering classification of rock masses for the design of tunnel support: Rock Mechanics, v. 6, p. 189–236, doi: 10.1007/BF01239496.

Biegel, R.L., Wang, W., Scholz, C.H., and Boitnott, B.N., 1992, Micromechanics of rock friction, 1. Effects of surface roughness on initial friction and slip hardening in Westerly granite: Journal of Geophysical Research, v. 97, B6, p. 8951–8964.

Bieniawski, Z.T., 1989, Engineering rock mass classification: New York, New York, John Wiley & Sons.

Bish, D.L., and Chipera, S.J., 1986, Mineralogy of drill holes J-13, UE-25A#1, and USW G-1 at Yucca Mountain, Nevada: Los Alamos, New Mexico, Los Alamos National Laboratory Data Report LA-10764-MS.

Boyd, P.J., Martin, R.J., III, and Price, R.H., 1994, An experimental comparison of laboratory techniques in determining bulk properties of tuffaceous rocks: Albuquerque, New Mexico, Sandia National Laboratories Data Report SAND92–0119.

Boyd, P.J., Price, R.H., Martin, R.J., III, and Noel, J.S., 1996a, Bulk and mechanical properties of the Paintbrush Tuff recovered from boreholes UE25 NRG-2, 2A, 2B, and 3: Albuquerque, New Mexico, Sandia National Laboratories Data Report SAND94–1902.

Boyd, P.J., Noel, J.S., Price, R.H., and Martin, R.J., III, 1996b, The effects of confining pressure on the strength and elastic properties of the Paintbrush Tuff recovered from boreholes USW NRG-6 and USW NRG-7/7A: Albuquerque, New Mexico, Sandia National Laboratories Data Report SAND95–1887.

Brace, W.F., Paulding, B.W., Jr., and Scholz, C.H., 1966, Dilatancy in the fracture of crystalline rocks: Journal of Geophysical Research, v. 71, p. 3939–3953.

Brailsford, A.D., and Major, K.G., 1964, The thermal conductivity of aggregates of several phases, including porous materials: British Journal of Applied Physics, v. 15, p. 313–319, doi: 10.1088/0508-3443/15/3/311.

Brechtel, C.E., Lin, M., Martin, E., and Kessel, D.S., 1995, Geotechnical characterization of the North Ramp of the Exploratory Studies Facility: Albuquerque, New Mexico, Sandia National Laboratories Data Report SAND95–0488/1 and /2.

Brodsky, N.S., Riggins, M., Connolly, J., and Ricci, P., 1997, Thermal expansion, thermal conductivity, and heat capacity measurements for boreholes UE25 NRG-4, UE25 NRG-5, USW NRG-6, and USW NRG-7/7A: Albuquerque, New Mexico, Sandia National Laboratories Data Report SAND95–1995.

Brown, E.T., ed., 1981, Rock characterization, testing and monitoring: ISRM suggested methods (published for the Commission on Testing Methods, International Society for Rock Mechanics): Elmsford, New York, Pergamon Press.

Brown, S.R., 1995, Simple mathematical model of a rough fracture: Journal of Geophysical Research, v. 100, B-4, p. 5941–5952, doi: 10.1029/94JB03262.

Buesch, D.C., Spengler, R.W., Moyer, T.C., and Geslin, J.K., 1996a, Proposed stratigraphic nomenclature and macroscopic identification of lithostratigraphic units of the Paintbrush Group exposed at Yucca Mountain, Nevada: Denver, Colorado, U.S. Geological Survey Open-File Report 94-469.

Christensen, R.J., Swanson, S.R., and Brown, W.S., 1974, Torsional shear measurements of the frictional properties of Westerly granite, *in* Advances in rock mechanics, Proceedings of the 3rd Congress of the International Society for Rock Mechanics: Washington, D.C., National Academy of Sciences, p. 221–225.

Chung, F.H., 1974, quantitative interpretation of X-ray diffraction patterns of mixtures, I. Matrix-flushing method for quantitative multicomponent analysis: Journal of Applied Crystallography, v. 7, p. 519–525, doi: 10.1107/S0021889874010375.

Connolly, J.R., and Nimick, F.B., 1990, Mineralogic and chemical data supporting heat capacity determination for tuffaceous rocks: Albuquerque, New Mexico, Sandia National Laboratories Data Report SAND88–0882.

CRWMS M&O (Civilian Radioactive Waste Management System, Management and Operating Contractor), 1995, Controlled design assumptions document: Las Vegas, Nevada, Civilian Radioactive Waste Management System, Management and Operating Contractor Report B00000000–01717–4600–00032 REV 02.

CRWMS M&O, 1996b, Characterization of the ESF Thermal Test Area: Las Vegas, Nevada, Civilian Radioactive Waste Management System, Management and Operating Contractor Report B00000000–01717–5705–00047 REV 01.

CRWMS M&O, 1996c, Synthesis of borehole and surface geophysical studies at Yucca Mountain, Nevada and vicinity. Volume II: Borehole geophysics: Las Vegas, Nevada, Civilian Radioactive Waste Management System, Management and Operating Contractor Report BAAA00000–01717–0200–00015 REV 00.

CRWMS M&O, 1997a, Evaluation of geotechnical monitoring data from the Exploratory Studies Facility, July 1995 to June 1996: Las Vegas, Nevada, Civilian Radioactive Waste Management System, Management and Operating Contractor Report BAB000000–01717–5705–00003 REV 01.

CRWMS M&O, 1997c, Determination of available volume for repository siting: Las Vegas, Nevada, Civilian Radioactive Waste Management System, Management and Operating Contractor Report BCA000000–01717–0200–00007 REV 00.

CRWMS M&O, 1997d, Yucca Mountain site geotechnical report: Las Vegas, Nevada, Civilian Radioactive Waste Management System, Management and Operating Contractor Report B00000000–01717–4600–00065 REV 01.

CRWMS M&O, 1997f, Confirmation of empirical design methodologies: TRW environmental safety systems: Las Vegas, Nevada, Civilian Radioactive Waste Management System, Management and Operating Contractor Report BABEE0000–01717–5705–00002 REV 00.

CRWMS M&O, 1997g, Ambient characterization of the Drift Scale Test block: Las Vegas, Nevada, Civilian Radioactive Waste Management System, Management and Operating Contractor Report BADD00000–01717–5705–00001 REV 01.

CRWMS M&O, 1997h, Single Heater Test status report: Las Vegas, Nevada, Civilian Radioactive Waste Management System, Management and Operating Contractor Report BAB000000–01717–5700–00002 REV 01.

CRWMS M&O, 2000, Drift degradation analysis/model report: Las Vegas, Nevada, Civilian Radioactive Waste Management System, Management and Operating Contractor Report ANL-EBS-000027 REV 00, February, 70 p., six attachments.

Deere, D.U., 1963, Technical description of rock cores for engineering purposes: Felsmechnik und Ingeniergeologie, v. 1, p. 16–22.

Geslin, J.K., Moyer, T.C., and Buesch, D.C., 1995, Summary of lithologic logging of new and existing boreholes at Yucca Mountain, Nevada, August 1993 to February 1994: U.S. Geological Survey Open-File Report 94–342.

Goktan, R.M., and Ayday, C., 1993, A suggested improvement to the Schmidt rebound hardness ISRM suggested method with particular reference to rock machineability: International Journal of Rock Mechanics and Mining Sciences & Geomechanics Abstracts, v. 30, no. 3, p. 321–322, doi: 10.1016/0148-9062(93)92733-7.

Halliday, D., and Resnick, R., 1974, Fundamentals of Physics: New York, J. Wiley and Sons.

Hardy, M.P., and Bauer, S.J., 1991, Drift design methodology and preliminary application for the Yucca Mountain site characterization project: Albuquerque, New Mexico, Sandia National Laboratories Date Report SAND89–0837.

Hoek, E., and Brown, E.T., 1988, The Hoek-Brown failure criterion—A 1988 update, *in* Proceedings of the Canadian Rock Mechanics Symposium: Ottawa, Canada, Mines Branch.

International Society for Rock Mechanics, 1981, Basic geotechnical description of rock masses: International Journal of Rock Mechanics and Mining Sciences & Geomechanics Abstracts, v. 18, no. 1, p. 85–110.

Jaeger, J.C., and Cook, N.G.W., 1979, Fundamentals of rock mechanics: London, England, Chapman and Hall.

Jung, J., Ryder, E.E., Boucheron, E.A., Dunn, E., Holland, J.F., and Miller, J.D., 1993, Design support analyses: North Ramp Design Package 2C: Albuquerque, New Mexico, Sandia National Laboratories Report TDIF 302218; DTN SNL01122093001.001.

Kicker, D.C., Martin, E.R., Brechtel, C.E., Stone, C.A., and Kessel, D.S., 1996, Geotechnical characterization of the Main Drift of the Exploratory Studies Facility: Albuquerque, New Mexico, Sandia National Laboratories Data Report SAND95–2183.

Kirsten, H.A.D., 1988, "Discussion on Q-system," *in* Kirkaldie, L., ed., Rock Classification Systems for Engineering Purposes Symposium held in Cincinnati, Ohio, 25 June 1987: Philadelphia, Pennsylvania, American Society for Testing Materials, p. 85–88.

Kutter, H.K., 1974, Rotary shear testing of rock joints, *in* Advances in rock mechanics, Proceedings of the 3rd Congress of the International Society for Rock Mechanics: Washington, D.C., National Academy of Sciences, p. 254–262.

Lappin, A.R., 1980, Preliminary thermal expansion screening data for tuffs: Albuquerque, New Mexico, Sandia National Laboratories Data Report SAND78–1147.

Lappin, A.R., and Nimick, F.B., 1985, Bulk and thermal properties of the functional tuffaceous beds in holes USW G-1, UE-25a#q, and USW G-2, Yucca Mountain, Nevada: Albuquerque, New Mexico, Sandia National Laboratories Data Report SAND82–1434.

Lappin, A.R., VanBuskirk, R.G., Enniss, D.O., Butters, S.W., Prater, F.M., Muller, C.B., and Bergosh, J.L., 1982, Thermal conductivity, bulk properties, and thermal stratigraphy of silicic tuffs from the upper portion of hole USW-G1, Yucca Mountain, Nye County, Nevada: Albuquerque, New Mexico, Sandia National Laboratories Data Report SAND81-1873: TIC 202545.

Lin, M., Hardy, M.P., and Bauer, S.J., 1993, Rock mass mechanical property estimations for the Yucca Mountain site characterization project: Albuquerque, New Mexico, Sandia National Laboratories Data Report SAND92–0450.

Majer, E.L., Feighner, M., Johnson, L., Daley, T., Karageorgi, E., Lee, K.H., Williams, K., and McEvilly, T., 1996, Surface geophysics. Volume I. Synthesis of borehole and surface geophysical studies at Yucca Mountain, Nevada and vicinity. Milestone B05M: Berkeley, California, Lawrence Livermore National Laboratory.

Martin, R.J., III, Price, R.H., Boyd, P.J., and Noel, J.S., 1994, Bulk and mechanical properties of the Paintbrush Tuff recovered from borehole USW NRG-6: Albuquerque, New Mexico, Sandia National Laboratories Data Report SAND93-4020.

Martin, R.J., III, Price, R.H., Boyd, P.J., and Noel, J.S., 1995, Bulk and mechanical properties of the Paintbrush Tuff recovered from borehole USW NRG-7/7A: Albuquerque, New Mexico, Sandia National Laboratories Data Report SAND94–1996.

Martin, R.J., III, Noel, J.S., Riggins, M.P., Boyd, J., and Price, R.H., 1997, Thermal expansion of the Paintbrush Tuff recovered from boreholes USW SD-12 at pressures to 30 MPa: Albuquerque, New Mexico, Sandia National Laboratories Data Report SAND95–1904.

McManus, T.T., 1996, Crystalline silica exposure monitoring—Yucca Mountain Exploratory Studies Facility, June 10: Las Vegas, Nevada, Environmental Health Services.

Michelis, P., and Brown, E.T., 1986, A yield equation for rock: Canadian Geotechnical Journal, v. 23, p. 9–17.

Nimick, F.B., 1989, Thermal-conductivity data for tuffs from the unsaturated zone at Yucca Mountain, Nevada: Albuquerque, New Mexico, Sandia National Laboratories Data Report SAND88–0624.

Nimick, F.B., 1990, The thermal conductivity of the Topopah Spring Member at Yucca Mountain, Nevada: Albuquerque, New Mexico, Sandia National Laboratories Data Report SAND86–0090.

Nimick, F.B., and Connolly, J.R., 1991, Calculation of heat capacities for tuffaceous units from the unsaturated zone at Yucca Mountain, Nevada: Albuquerque, New Mexico, Sandia National Laboratories Data Report SAND88–3050.

Nimick, F.B., and Lappin, A.R., 1985, Thermal conductivity of silicic tuffs from Yucca Mountain and Rainier Mesa, Nye County, Nevada: Albuquerque, New Mexico, Sandia National Laboratories Data Report SAND83–1711J.

Nimick, F.B., Price, R.H., Van Buskirk, R.G., and Goodell, J.R., 1985, Uniaxial and triaxial compression test series on Topopah Spring Tuff from USW G-4, Yucca Mountain, Nevada: Albuquerque, New Mexico, Sandia National Laboratories Data Report SAND84–1101.

Nimick, F.B., Shephard, L.E., and Blejwas, T.E., 1988, Preliminary evaluation of the exploratory shaft representativeness for the Nevada Nuclear Waste Storage Investigations Project: Albuquerque, New Mexico, Sandia National Laboratories Data Report SAND87–1685.

Olsson, W.A., 1982, Effects of elevated temperature and pore pressure on the mechanical behavior of Bullfrog Tuff: Albuquerque, New Mexico, Sandia National Laboratories Data Report SAND81–1664.

Olsson, W.A., 1987, Rock joint compliance studies: Albuquerque, New Mexico, Sandia National Laboratories Data Report SAND86–0177.

Olsson, W.A., 1988, Compliance and strength of artificial joints in Topopah Spring Tuff: Albuquerque, New Mexico, Sandia National Laboratories Data Report SAND88–0660.

Olsson, W.A., and Brown, S.R., 1993, Hydromechanical response of a fracture undergoing compression and shear: International Journal of Rock Mechanics, Mining Sciences & Geomechanics Abstracts, v. 30, p. 845–851, doi: 10.1016/0148-9062(93)90034-B.

Olsson, W.A., and Brown, S.R., 1994, Mechanical properties of seven fractures from drillholes NRG-4 and NRG-6 at Yucca Mountain, Nevada: Albuquerque, New Mexico, Sandia National Laboratories Data Report SAND94–1995.

Olsson, W.A., and Brown, S.R., 1995, mechanical properties of fractures from drillholes UE25-NRG-4, USW-NRG-6, USW-NRG-7, and USW-SD-9 at Yucca Mountain, Nevada: Albuquerque, New Mexico, Sandia National Laboratories Data Report SAND95–1736.

Olsson, W.A., and Brown, S.R., 1997, Mechanical properties of fractures from drillholes, UE25-NRG-4, USW-NRG-6, USW-NRG-7, USW-SD-9 at Yucca Mountain, Nevada: Albuquerque, New Mexico, Sandia National Laboratories Data Report SAND95–1736.

Olsson, W.A., and Jones, A.K., 1980, Rock mechanics properties of volcanic rock units from the Nevada Test Site: Albuquerque, New Mexico, Sandia National Laboratories Data Report SAND80–1453.

Ortiz, T.S., Williams, R.L., Nimick, F.B., Whittet, B.C., and South, D.L., 1985, A three-dimensional model of reference thermal/mechanical and hydrological stratigraphy at Yucca Mountain, Southern Nevada: Albuquerque, New Mexico, Sandia National Laboratories Data Report SAND84–1076.

Palmstrom, A., 1996, Characterizing rock masses by the RMi for use in practical rock engineering. Part 1: The development of the rock mass index (RMi): Tunneling and Underground Space Technology, v. 11, no. 2.

Papike, J.J., and Cameron, M., 1976, Crystal chemistry silicate minerals of geophysical interest: Reviews of Geophysics and Space Physics, v. 14, no. 1, p. 37–80.

Price, R.H., 1983, Analysis of rock mechanics properties of volcanic tuff units from Yucca Mountain, Nevada Test Site: Albuquerque, New Mexico, Sandia National Laboratories Data Report SAND82–1315.

Price, R.H., 1986, Effects of sample size on the mechanical behavior of Topopah Spring Tuff: Albuquerque, New Mexico, Sandia National Laboratories Data Report SAND85–0709.

Price, R.H., and Bauer, S.J., 1985, Analysis of the elastic and strength properties of Yucca Mountain Tuff, Nevada, in Ashworth, E., ed., Proceedings of the 26th U.S. Symposium on Rock Mechanics, Rapid City, South Dakota: Boston, Massachusetts, A.A. Balkema, p. 89–96.

Price, R.H., and Jones, A.K., 1982, Unixial and triaxial compression tests series on Calico Hills Tuff: Albuquerque, New Mexico, Sandia National Laboratories Data Report SAND82–1314.

Price, R.H., and Nimick, K.G., 1982, Uniaxial compression test series on Tram Tuff: Albuquerque, New Mexico, Sandia National Laboratories Data Report SAND82–1055.

Price, R.H., Jones, A.K., and Nimick, K.G., 1982a, Uniaxial and triaxial compression test series on Bullfrog Tuff: Albuquerque, New Mexico, Sandia National Laboratories Data Report SAND82–0481.

Price, R.H., Nimick, K.G., and Zirzow, J.A., 1982b, Uniaxial and triaxial compression test series on Topopah Spring Tuff: Albuquerque, New Mexico, Sandia National Laboratories Data Report SAND82–1723.

Price, R.H., Spence, S.J., and Jones, A.K., 1984, Uniaxial compression test series on Topopah Spring Tuff from USW GU-3, Yucca Mountain, Southern Nevada: Albuquerque, New Mexico, Sandia National Laboratories Data Report SAND83–1646.

Price, R.H., Nimick, F.B., Connolly, J.R., Keil, K., Schwartz, B.M., and Spence, S.J., 1985, Preliminary characterization of the petrologic, bulk, and mechanical properties of a lithophysal zone within the Topopah Spring Member of the Paintbrush Tuff: Albuquerque, New Mexico, Sandia National Laboratories Data Report SAND84–0860.

Price, R.H., Martin, R.J., III, and Boyd, P.J., 1993, Characterization of porosity in support of mechanical property analysis: High-level radioactive waste management, in Proceedings of the Fourth Annual International Conference, Las Vegas, Nevada, v. 2, p. 1847–1853.

Rautman, C.A., 1995, Preliminary geostatistical modeling of thermal conductivity for a cross section of Yucca Mountain, Nevada: Albuquerque, New Mexico, Sandia National Laboratories Data Report SAND94–2283.

Rautman, C.A., and McKenna, S.A., 1997, Three-dimensional hydrological and thermal property models of Yucca Mountain, Nevada: Albuquerque, New Mexico, Sandia National Laboratories Data Report SAND97–1730.

Roberts, S.K., and Viani, B.E., 1997, Mineral abundances for samples from six chemistry (Seamist) boreholes in the Drift Scale Test Area (DST) of the ESF: Yucca Mountain Site Characterization Project, Level 4 Milestone SP9510M4: Livermore, California, Lawrence Livermore National Laboratory.

Sass, J.H., Lachenbruch, A.H., Dudley, W.W., Jr., Priest, S.S., and Munroe, R.J., 1988, Temperature, thermal conductivity, and heat flow near Yucca Mountain, Nye County, Nevada: Soume tectonic and hydrologic implications: U.S. Geological Survey Open-File Report 87-649, 118 p.

Scholz, C.H., 1968, Microfracturing and the inelastic deformation of rock in compression: Journal of Geophysical Research, v. 73, p. 1417.

Schwartz, B.M., and Chocas, C.S., 1992, Linear thermal expansion data for tuffs from the unsaturated zone at Yucca Mountain, Nevada: Albuquerque, New Mexico, Sandia National Laboratories Data Report SAND88–1581

Serafim, J.L., and Pereira, J.P., 1983, Consideration of the geomechanical classification of Bieniawski, in Proceedings, International Symposium on Engineering Geology and Underground Construction, held in Lisbon, Portugal: International Association of Engineering and Geology, p. 1133–1144.

Snyder, R.L., and Bish, D.L., 1989, Quantitative analysis, in Bish, D.L., and Post, J.E., eds., Modern powder diffraction: Washington, D.C., Mineralogical Society of America Reviews in Mineralogy, v. 20, p. 101–144.

Spengler, R.W., and Fox, K.F., Jr., 1989, Stratigraphic and structural framework of Yucca Mountain, Nevada: Radioactive Waste Management and the Nuclear Fuel Cycle, v. 13, p. 21–36.

SNL (Sandia National Laboratories) TP-202, Measurement of thermal conductivity of geologic samples by the guarded-heat-flow meter method: Albuquerque, New Mexico, Sandia National Laboratories.

SNL TP-203, Measurement of thermal expansion of geologic supplies using a push rod dilatometer: Cambridge, Massachusetts, Holometrix, Inc.

SNL TP-229, Bulk properties determinations of tuffaceous rocks: Dry bulk density, saturated bulk density, average grain density, and porosity: Albuquerque, New Mexico, Sandia National Laboratories.

SNL, 1995, Evaluation of geotechnical monitoring data from the ESF North Ramp Starter Tunnel, April 1994 to June 1995: Albuquerque, New Mexico, Sandia National Laboratories Data Report SAND95–1675.

SNL, 1997a, Hydraulic fracturing stress measurements in test hole ESF-AOD-HDFR#1, Thermal Test Facility, Exploratory Studies Facility at Yucca Mountain: Albuquerque, New Mexico, Sandia National Laboratories.

SNL, 1997b, Evaluation and comparative analysis of Single Heater Test thermal and thermomechanical data: Third quarter results (8/26/96–5/31/97). TDIF 306195. Level 4 Milestones SP9261M4, SP9271M4, and SP9268M4: Albuquerque, New Mexico, Sandia National Laboratories.

Stock, J.M., Healy, J.H., and Hickman, S.H., 1984, Report on televiewer log and stress measurements in core hole USW G-2, Nevada Test Site, October–November 1982: U.S. Geological Survey Open-File Report 84–172.

Sweetkind, D.S., and Williams-Stroud, S.C., 1996, Characteristics of fractures at Yucca Mountain, Nevada: Synthesis Report: Denver, Colorado,

U.S. Geological Survey, Administrative Report to the U.S. Department of Energy.

Thompson, A.B., and Wennemer, M., 1979, Heat capacities and inversions in tridymite, cristobalite, and tridymite-cristobalite mixed phases: The American Mineralogist, v. 64, p. 1018–1026.

Tullis, T.E., and Weeks, J.D., 1986, Constitutive behavior and stability of frictional sliding of granite: Pageoph, v. 124, p. 383–414, doi: 10.1007/BF00877209.

University of Toronto Rock Engineering Group, 1992, UNWEDGE Version 2.2: Ontario, Canada, University of Toronto.

Vaniman, D.T., Bish, D.L., Chipera, S.J., Carlos, B.A., and Guthrie, G.D., 1996, Summary and synthesis report on mineralogy and petrology studies for the Yucca Mountain Site Characterization Project. Volume I: Chemistry and mineralogy of the transport environment at Yucca Mountain: Los Alamos, New Mexico, Los Alamos National Laboratory Milestone 3665.

Viani, B., and Roberts, S., 1996, Determination of mineral abundances in core samples from the Exploratory Studies Facility using X-ray diffraction. Analysis report supporting Yucca Mountain Project, TDIF 305632, Mineral Abundances for Temperature Samples from the Thermal Area of the ESF: Livermore, California, Lawrence Livermore National Laboratory.

Weast, R.C., 1974, CRC handbook of chemistry and physics (55th edition): Cleveland, Ohio, CRC Press, p. F-109, F-112.

Weeks, J.D., and Tullis, T.E., 1985, Frictional sliding of dolomite: A variation in constitutive behavior: Journal of Geophysical Research, v. 90, p. 7821–7826.

Woodside, W., and Messmer, J.H., 1961, Thermal conductivity of porous media: I. Unconsolidated sands: Journal of Applied Physics, v. 32, no. 9, p. 1688–1699, doi: 10.1063/1.1728419.

Xu, S., and de Frietas, M.H., 1988, Use of rotary shear box for testing the shear strength of rock joints: Geotechnique, v. 38, p. 301–309.

YMP (Yucca Mountain Project), 1995, Reference information base. Report YMP/93-02, Rev. 3, ICN 1: Las Vegas, Nevada, Yucca Mountain Site Characterization Office.

Yoshioka, N., and Scholz, C.H., 1989, Elastic properties of contacting surfaces under normal and shear loads: 2. Comparison of theory with experiment: Journal of Geophysical Research, v. 94, p. 17,691–17,700.

Young, R.A., 1993, Introduction to the Rietveld method. The Rietveld method: International Union of Crystallography, Monographs on Crystallography, v. 5, p. 1–38.

Yudhbir, L.W., and Prinzl, F., 1983, An empirical failure criterion for rock masses, *in* Proceedings of the 5th International Congress on Rock Mechanics, Melbourne, Australia: Rotterdam, Netherlands, A.A. Balkema for the International Society for Rock mechanics, p. B1–B8.

Zimmerman, R.M., and Finley, R.E., 1987, Summary of geomechanical measurements taken in and around the G-tunnel Underground Facility, NTS: Albuquerque, New Mexico, Sandia National Laboratories Data Report SAND86–1015.

Zoback, M.D., and Healy, J.H., 1984, Friction, faulting, and in situ stress: Annales Geophysicae, v. 2, no. 6, p. 689–698.

MANUSCRIPT ACCEPTED BY THE SOCIETY 29 DECEMBER 2005

Geological Society of America
Special Paper 408
2006

Chapter 4.1

The hydrology of tuffs

Rebecca C. Smyth*

Bureau of Economic Geology, Jackson School of Geosciences, The University of Texas at Austin, Austin, Texas 78713, USA

John M. Sharp Jr.

Department of Geological Sciences, Jackson School of Geosciences, The University of Texas at Austin, Austin, Texas 78713, USA

INTRODUCTION

The definition of *tuff* given by the American Geological Institute is: "A rock formed of compacted volcanic fragments, generally smaller than 4 mm in diameter." However, in practice and in the context of this book, the term *tuff* applies to the variety of deposits originating from volcanic eruptions of high-silica magmas. These deposits are dominantly referred to as *air-fall* or *ash-flow* tuffs. *Bedded tuff* is another term used inconsistently in hydrologic studies to refer to both air-fall tuffs and reworked tuffaceous sediments. Ash-flow tuffs are commonly separated into three categories: nonwelded, partially welded, and densely welded (Smith, 1960). Welding is dependent upon compaction of the tuff components upon or soon after deposition and the temperature of particles and gases upon deposition (Ross and Smith, 1961). Nonwelded tuffs have low bulk density and are friable; therefore, they have high matrix porosity and permeability. Densely welded tuffs have higher bulk density and, therefore, low matrix porosity and permeability, but are brittle and so are more highly fractured.

Interest in the hydrologic properties of tuffs has grown within the past ~35 yr in the United States for two reasons. Scarce water resources in the southwestern United States prompted groundwater hydrologists to consider the widespread fractured igneous rocks of that area as potential aquifers. Secondly, the need arose to understand groundwater flow systems at Yucca Mountain, Nevada. Yucca Mountain is the only site in the United States being considered for permanent disposal of the nation's high-level radioactive waste. The enormous physical and chemical heterogeneity of tuff has fueled debate over how to characterize

the hydrologic properties required to predict groundwater movement and potential radionuclide migration at Yucca Mountain. U.S. scientists have done a thorough job of collecting and analyzing data to accomplish this objective, primarily through studies conducted on the tuffs at Yucca Mountain.

Characterizing the hydrologic properties of tuffs at Yucca Mountain is most complex, in part, because of the great thickness of tuffs present at this site. Yucca Mountain, located on the western edge of the Nevada Test Site in southwestern Nevada, USA, is a 1- to 2-km-thick accumulation of volcanic rock erupted from the Claim Canyon caldera of the Timber Mountain–Oasis Valley caldera complex (between 14 and 11.3 m.y. ago). The rocks are rhyolite to quartz latite air-fall and ash-flow (ignimbrite) tuffs, tuffaceous sediments, and surge deposits (Lipman et al., 1972; Byers et al., 1976). The horizon in which waste is to be isolated is between 200 and 400 m deep within the 12.7 Ma Topopah Spring Member of the Paintbrush Tuff. Prevailing thought is that this is a good site for a waste repository because it is within the unsaturated zone, ~300–400 m above the regional water table. Long-term integrity of the repository will depend upon the physical hydrologic properties of the Topopah Spring, a densely welded ash-flow tuff, and overlying partially welded ash-flow (Yucca and Pah Canyon Members) and air-fall tuffs. The Calico Hills Tuff that underlies the repository horizon is postulated to buffer radionuclide migration should (or more pessimistically *when*) a release occurs.

"Analog" (analogous to Yucca Mountain) studies of tuff hydrologic properties have been conducted on the Santana Tuff in Trans-Pecos Texas, the Apache Leap Tuff in south-central Arizona, and the Nopal Tuff in Chihuahua, Mexico. The Santana Tuff is a rhyolitic, variably welded tuff that erupted from the Sierra Rica caldera complex in northern Chihuahua, Mexico,

*E-mail: rebecca.smyth@beg.utexas.edu.

Smyth, R.C., and Sharp, J.M., Jr., 2006, The hydrology of tuffs, *in* Heiken, G., ed., Tuffs—Their properties, uses, hydrology, and resources: Geological Society of America Special Paper 408, p. 91–111, doi: 10.1130/2006.2408(4.1). For permission to copy, contact editing@geosociety.org.

ca. 28 Ma and crops out within a 125 km^2 area of Trans-Pecos Texas (Chuchula, 1981; Gregory, 1981; Henry et al., 1986). The ca. 20 Ma Apache Leap Tuff lies within the upper, slightly welded portion of a tuff formation located near Superior, Arizona (Peterson, 1962). The Tertiary-aged, rhyolitic Nopal Tuff lies within the Peña Blanca Uranium District, Chihuahua, Mexico. Researchers have studied hydrologic properties of the Bandelier Tuff in New Mexico as an analog to Yucca Mountain and because of local interests in waste containment (i.e., Neeper and Gilkeson, 1996). The Bandelier Tuff is a rhyolitic tuff that erupted to form the Valles caldera in Jemez volcanic field in two phases, 1.6 and 1.2 m.y. ago (Heiken et al., 1990).

The andesitic ignimbrites within a Quaternary volcanic sequence in Valle Central, Costa Rica, are of interest to hydrologists because they help serve rapidly expanding urban and industrial (coffee plantations) water supply needs (Foster et al., 1985). In Indonesia, tuffaceous rocks host geothermal power and water supply resources.

In this chapter, we present hydrologic properties of tuffs in two categories: laboratory data and in situ data. Measurement of tuff matrix hydrologic properties, such as bulk density, porosity, pore-size distribution, permeability, and saturated hydraulic conductivity, are usually conducted in a laboratory on core samples. In situ tests are carried out within vertical boreholes and wells and include (1) aquifer pumping and recovery tests, (2) packer-injection or slug tests using air or water, (3) radioactive tracer and temperature borehole surveys, and (4) geophysical logging. A relatively new method of in situ testing is being utilized in a tunnel in Yucca Mountain, where it is possible to more accurately describe the tuffs and to view hydrologic processes in action. Measurements of fracture spacing or density, aperture, fracture surface roughness, and connectivity can be conducted in outcrop or on cores in a laboratory. Estimation of tuff hydrologic properties has also been performed through inverse numerical and stochastic modeling. We preface the data presentation with a review of basic hydrologic concepts and definition of properties reported in the tables. We end the chapter with an interpretation of the wide range of hydrologic properties that have been measured in tuffs.

HYDROLOGIC PROPERTIES

Hydrologic properties are those physical parameters that control the flow and distribution of water and solutes in the subsurface. These are, perhaps, defined more precisely as hydraulic or hydrogeologic parameters. These properties have great consequence to plant growth, rates of weathering, slope stability, groundwater recharge, and various geotechnical considerations, including analysis of tuff systems as repositories for wastes. The key properties are porosity, which limits the amount of water a rock or soil can hold, and permeability, which is the ease with which a porous medium can transmit a fluid.

Porosity is the percentage of void space in a rock or soil (the porous medium). Porosity can consist of intergranular voids in the rock matrix or voids resulting from open fractures, dissolution features, or vugs. In rocks described as having dual-porosity systems, porosity occurs in the rock matrix and in fractures. Matrix and fracture porosities are referred to, respectively, as primary porosity, which develops when the tuff is deposited as ash flows or ash falls, and secondary porosity, which develops after the tuff has been deposited (fractures are the most critical secondary porosity in tuffs). Primary porosity in volcanic rocks includes vesicles (gas bubbles in pumice or scoria) or is created when gases, trapped between particles soon after deposition, escape and leave a void space. Secondary mineral growth can result in crystal-lined cavities called *lithophysae*. Because tuffs contain varying amounts of pumice fragments, vesicles, and lithophysae, and are welded to varying degrees, they can have porosities ranging from <5% to >50%.

Porosity (n) is calculated by:

$$n = V_V / V_T, \qquad (4.1)$$

where V_V = volume of void space, and V_T = total volume of the rock. Porosity is usually calculated by measuring the mass of a porous solid of known volume (m_s), which has lost all its water by oven drying. The estimated density of the solid phase (ρ_s), which is mostly silica and feldspar in tuffs, ranges from 2.56 to 2.65 g/cm^3, so equation 4.1 can be rewritten

$$n = \frac{V_T - \dfrac{m_s}{\rho_s}}{V_T}. \qquad (4.2)$$

Less commonly, porosity is measured using mercury-injection porosimetry (e.g., Smyth-Boulton, 1995), which also gives a pore-size distribution, or argon gas sorption.

When the pore spaces are filled with water, the tuff is termed saturated. Two moisture content definitions are used—the volumetric moisture content (θ) and the gravimetric moisture content (w). They are defined as follows:

$$\theta = V_W / V_V \qquad (4.3)$$

and

$$w = m_W / m_S, \qquad (4.4)$$

where V_W is the volume of water in the sample, and m_W is the mass of water in the sample. Thus, θ can vary between zero and n. The mass of water is calculated by weighing a sample of known volume before and after oven drying. The volumetric water content is a function of the gravimetric water content, or

$$\theta = w \cdot m_S / V \rho_W. \qquad (4.5)$$

In general, tuffs that have higher primary porosity also have higher permeability. Deviations from this relationship arise when (1) pores are filled with secondary mineralization (i.e., zeolitization in tuffs) or (2) pores lack interconnectedness, thereby impeding water flow. Pumice fragments, vesicles, and lithophysae in tuffs are usually only connected by fractures, so the increased porosity associated with these features may not necessarily result in increased permeability of the tuff. Permeability is the ease with which fluid can flow through a porous medium. Units measuring permeability can be confusing, especially to those not trained in hydrogeology. For example, intrinsic permeability (k) is reported in units of length squared (L^2) or in darcies. (A darcy is a discharge of 1 cm^3/s of a fluid with a viscosity of 1 centipoise through an area of 1 cm^2 for a hydraulic gradient of 1 atm/cm. A millidarcy equals 10^{-3} darcy, and one darcy equals 9.87×10^{-9} cm^2.)

Intrinsic permeability is a property of the porous medium and is independent of the properties of the fluid. For this reason, intrinsic permeability is uniformly used in the petroleum industry, which routinely deals with several fluids—natural gas, oil, and water. Hydraulic conductivity (K) is derived from Darcy's law (Darcy, 1856) as the discharge (Q) per unit area (A) per unit hydraulic gradient normal to the area (∇h),

$$K = \frac{Q}{\nabla h \cdot A}. \tag{4.6}$$

Hydraulic conductivity (K) is reported in units of length over time ($L\,t^{-1}$) or in old American units of gallons per day per square foot (gpd/ft²). The relationship between intrinsic permeability (k) and hydraulic conductivity (K) was defined by Hubbert (1956) as

$$K = k\frac{\rho_w g}{\mu}, \tag{4.7}$$

where ρ_w is fluid density ($M\,L^{-3}$), g is gravitational acceleration ($L\,t^{-2}$), and μ is the dynamic viscosity.

Another commonly reported unit of permeability is transmissivity (T), which is reported in units of length squared over time ($L^2\,t^{-1}$) or in old American units of gallons per day per foot (gpd/ft). Transmissivity is the discharge through a unit strip of aquifer (one unit wide and over the entire saturated thickness) per unit hydraulic gradient normal to that strip. In the strictest sense, transmissivity applies only to horizontal flow. Thus, transmissivity is the horizontal hydraulic conductivity integrated over the saturated thickness of the aquifer (b) or

$$T = \int_{o}^{b} K_{horizontal}(z)\,dz. \tag{4.8}$$

If the hydraulic conductivity is homogeneous, then T and K are related by

$$T = Kb. \tag{4.9}$$

Permeability and hydraulic conductivity are commonly measured in the laboratory on samples of rock matrix or granular soils (e.g., ASTM, 2000; Sharp et al., 1994) through which water or air is flowing and in the field by a variety of pumping, packer, and piezometer tests (e.g., Butler, 1998; Sara, 2003). Inverse numerical modeling can also provide estimations of permeability.

Our search of the literature plus our studies show intrinsic permeabilities reported in cm², m², or darcies; hydraulic conductivities reported in cm/s, m/d, ft/d, and gpd/ft²; and transmissivities reported in m²/d, ft²/min, and gpd/ft. Finally, we need to note that the hydraulic conductivity, but not the intrinsic permeability, is dependent upon the amount of fluid present—the permeability of a tuff to water is greatest when the rock is saturated with water. Consequently, our hydraulic conductivity data assume saturated conditions. In the data tables of tuff hydrologic properties, we list the original reporting units, but we have converted all intrinsic permeability and saturated hydraulic conductivity measurements into m² and m/d, respectively, for ease of comparison of values between data sets.

Tuff Hydrologic Data

The hydrologic data we present are highly variable because tuff is a very heterogeneous type of rock; porosity ranges from <5% to >50%, and permeability ranges over 11 orders of magnitude. Heterogeneities in tuffaceous rocks depend upon the primary characteristics of: (1) percentage of phenocrysts and rock fragments, which are dependent upon magma composition and terrain type through which magma rose prior to eruption and over which flows passed after eruption, and (2) degree of welding (sintering of glass upon or soon after deposition), cooling rate, and amount of compaction. For example, subsequently erupted layers of ash deposited upon older layers would trap expanding gas (possibly resulting in increased matrix porosity), collapse pumice fragments, flatten glass shards, and retard cooling, which would increase the degree of welding and result in more closely spaced cooling fractures. Secondary processes that influence hydrologic properties of tuffs are (1) postdepositional mineralization, which includes devitrification of glass shards and crystallization of vapor-phase minerals and zeolites, and (2) tectonic stresses resulting in various scales of fractures and faults. The hydrologic properties of tuffs are controlled by complex geological processes acting upon the rocks over long periods of time.

We present two categories of tuff hydrologic data: laboratory matrix data (Table 4.1.1), and in situ well test data (Table 4.1.2). Tables 4.1.1 and 4.1.2 contain data from 20 and 9 references, respectively. Data in both tables are primarily from measurements of Yucca Mountain tuffs. Limited laboratory and well test data are also available for tuffs from other locations, including: (1) Santana Tuff, Trans-Pecos Texas, (2) Apache Leap Tuff, near Superior, Arizona, (3) Bandelier Tuff, Parajito Plateau, New Mexico, (4) Nopal Tuff, Chihuahua, Mexico, (5) tuffs in Valle Central, Costa Rica, and (6) tuffs in the Bandung Basin, Indonesia. Many more measurements of tuff matrix hydrologic properties can be found in the literature, primarily because they are easier and less costly to perform; especially compared to in situ testing, which requires drilling up to 1000-ft-deep wells or tunneling into the side of a mountain. Determining fracture properties, such as fracture spacing, fracture orientation, and fracture aperture, is essential to fully understanding the hydrologic properties of tuffs;

TABLE 4.1.1.

Locations/Units Sampled	Degree of welding	Matrix porosity (%)	Water content (%)	Dry bulk density (g/cm³)	Matrix permeability (reported units)	Matrix permeability (m²)	Hydraulic conductivity (m/sec)	Hydraulic conductivity (m/day)	Source
A. Yucca Mtn. Boreholes USW GU-3/G-3 and G-4 (Avg. of 160 samples)		17.59		2.06	(md)				Anderson (1994)
(Range)		2.2–40.9		1.39–2.45	2.0E-5–200			1.7E-2 to 1.7E+5	
B. Yucca Mtn. Test Well UE-25a#1 Topopah Spring Tuff	welded					8.05E-19 / 9.20E-21	7.00E-12 / 8.00E-14	6.05E-07 / 6.91E-09	Anderson (1981)
C. Yucca Mountain devitrified ignimbrite	moderate to dense	9.9		2.24		6.09E-18 / 1.44E-15 / 6.18E-18 / 8.76E-19	3.81E-08 / 1.25E-08 / 5.37E-11 / 7.62E-12	4.57E-06 / 1.08E-03 / 4.64E-06 / 6.58E-07	Bandurraga and Bodvarsson (1997) Peters et al. (1984)
bedded, vitric	non to partial	47		1.28		1.80E-14 / 3.72E-15 / 6.09E-14 / 9.92E-14 / 3.61E-13 / 2.13E-15	1.64E-07 / 3.31E-08 / 5.35E-07 / 8.75E-07 / 3.17E-06 / 1.85E-08	1.35E-02 / 2.79E-03 / 4.57E-02 / 7.45E-02 / 2.71E-01 / 1.60E-03	
devitrified, lithophysae > 10% by volume	moderate to dense	3.1		2.49		2.25E-17 / 2.59E-17 / 1.11E-16 / 9.89E-20	2.00E-10 / 2.26E-10 / 9.62E-10 / 8.60E-13	1.69E-05 / 1.95E-05 / 8.31E-05 / 7.43E-08	
devitrified, lithophysae > 10% by volume	moderate to dense	10		2.3		8.57E-17 / 1.92E-16 / 4.51E-18 / 8.45E-18 / 1.12E-18	7.62E-10 / 1.67E-09 / 3.98E-11 / 7.52E-11 / 9.77E-12	6.44E-05 / 1.44E-04 / 3.39E-06 / 6.35E-06 / 8.44E-07	
vitrophyre	dense	6.0		2.27		5.64E-17 / 5.64E-18 / 3.52E-15 / 8.42E-17 / 1.52E-18	4.98E-10 / 4.96E-11 / 3.06E-08 / 7.32E-10 / 1.32E-11	4.23E-05 / 4.23E-06 / 2.64E-03 / 6.32E-05 / 1.14E-06	
bedded, reworked, vitric	nonwelded	38		1.47		1.80E-18 / 6.20E-14 / 6.34E-14 / 9.09E-15	1.60E-05 / 2.30E-09 / 5.51E-07 / 7.90E-08	1.35E-06 / 4.66E-02 / 4.76E-02 / 6.83E-03	
bedded, reworked, zeolitic	nonwelded	30		1.62		6.76E-18 / 6.44E-19 / 4.83E-19	5.37E-09 / 5.60E-12 / 4.20E-12	5.08E-06 / 4.84E-07 / 3.63E-07	

(continued)

TABLE 4.1.1. (continued)

Locations/Units Sampled	Degree of welding	Matrix porosity (%)	Water content (%)	Dry bulk density (g/cm³)	Matrix permeability (reported units)	Matrix permeability (m²)	Hydraulic conductivity (m/sec)	Hydraulic conductivity (m/day)	Source
C. Yucca Mountain (*continued*) welded, bedded, zeolitic	nonwelded	27		1.70		5.07E-18 9.47E-18 8.67E-18 1.67E-18	4.54E-11 8.35E-11 7.54E-11 1.45E-11	3.81E-06 7.11E-06 6.51E-06 1.25E-06	
devitrified	moderate	29		1.82		2.14E-15 1.06E-17 1.08E-17 1.00E-17 1.02E-16 3.50E-16	1.89E-08 2.84E-10 9.35E-11 8.70E-11 8.84E-10 3.04E-09	1.61E-03 7.96E-06 8.08E-06 7.52E-06 7.64E-05 2.63E-04	
bedded, reworked, zeolitized	non to partial	24		1.72		1.08E-17	9.41E-11	8.13E-06	
devitrified	moderate to dense	26		1.95		4.37E-16	3.80E-09	3.28E-04	Dailey et al. (1987)
D. Yucca Mountain Topopah Springs (1 sample)	welded	8.45		2.32	(md) 9	8.96E-15		7.51E-03	Daniels et al. (1982)
E. Yucca Mountain Topopah Springs, Calico Hills, and Bullfrog (10 samples)	partial to dense	7–40 Avg. = 25				**1.00E-19 to 2.50E-17**		8.47E-08 to 2.12E-05	
F. Yucca Mountain Tuffs	variably welded	14–51		1.4–1.9	(md) 0.01–362	9.97E-18 to 3.61E-13		8.36E-06 to 3.03E-01	Flint and Flint (1990)
G. Yucca Mountain Tuffs core from 31 boreholes									Flint (1998)
Paintbrush Group Tiva Canyon Tuff									
1. Unit CCR (9 samples)	vitrophyre	6.2 ± 2	4.6 ± 2			1.725E-19	1.5E-12 ± 2.1E-12	1.30E-07	
2. Unit CUC (101 samples)	partially	25.3 ± 6	9.8 ± 5			4.485E-15	3.9E-08 ± 1.4E-07	3.37E-03	
3. Unit CUL (98 samples)	to	16.4 ± 6	9.4 ± 2			6.555E-17	5.7E-10 ± 1.2E-07	4.92E-05	
4. Unit CW (599 samples)	moderately	8.2 ± 3	6.4 ± 3			4.37E-19	3.8E-12 ± 4.1E-09	3.28E-07	
5. Unit CMW (90 samples)	welded	20.3 ± 5	18.5 ± 6			1.012E-18	8.8E-12 ± 1.0E-11	7.60E-07	
6. Unit CNW (101 samples)	nonwelded	38.7 ± 7	25.9 ± 8			2.99E-14	2.6E-07 ± 2.2E-06	2.25E-02	

(continued)

TABLE 4.1.1. (continued)

Locations/Units Sampled	Degree of welding	Matrix porosity (%)	Water content (%)	Dry bulk density (g/cm³)	Matrix permeability (reported units)	Matrix permeability (m²)	Hydraulic conductivity (m/sec)	Hydraulic conductivity (m/day)	Source
G. Yucca Mountain Tuffs (*continued*)									
Yucca Mountain Tuff									
7. Unit TPY (43 samples)	nonwelded	25.4 ± 8	16.3 ± 5			1.955E-15	1.7E-08 ± 7.4E-07	1.47E-03	
Pah Canyon									
8. Unit TPP (164 samples)	nonwelded	49.9 ± 4	17.8 ± 6			4.14E-13	3.6E-06 ± 1.2E-06	3.11E-01	
Topopah Springs Tuff									
9. Unit TC (66 samples)	densely welded	5.4 ± 4	3.4 ± 2			7.13E-20	6.2E-13 ± 7.1E-10	5.36E-08	
10. Unit TR (439 samples)	partially welded	15.7 ± 3	7.8 ± 2			4.485E-17	3.9E-10 ± 1.0E-09	3.37E-05	
11. Unit TUL (455 samples)	partially welded	15.4 ± 3	10.8 ± 2			2.645E-17	2.3E-10 ± 5.2E-10	1.99E-05	
12. Unit TMN (266 samples)	partially welded	11.0 ± 2	9.3 ± 2			1.725E-18	1.5E-11 ± 2.3E-11	1.30E-06	
13. Unit TLL (453 samples)	partially welded	13.0 ± 3	10.1 ± 2			8.05E-18	7.0E-11 ± 4.5E-10	6.05E-06	
14. Unit TM2 (225 samples)	partially welded	11.2 ± 3	9.5 ± 3			2.07E-18	1.8E-11 ± 2.1E-09	1.56E-06	
15. Unit TM1 (102 samples)	partially welded	9.4 ± 2	8.1 ± 2			5.52E-19	4.8E-12 ± 5.8E-12	4.15E-07	
16. Unit PV3 (89 samples)	densely welded	3.6 ± 4	3.4 ± 4			1.725E-20	1.5E-13 ± 2.1E-12	1.30E-08	
17. Unit PV2 (39 samples)	nonwelded	17.3 ± 11	14.8 ± 11			8.51E-18	7.4E-11 ± 2.0E-07	6.39E-06	
Calico Hills Formation									
18. Unit CHV (69 samples)	nonwelded, vitric	34.5 ± 3	16.8 ± 8			2.415E-14	2.1E-07 ± 5.5E-07	1.81E-02	
19. Unit CHZ (293 samples)	nonwelded, zeolitic	33.1 ± 4	32.0 ± 4			1.265E-17	1.1E-10 ± 1.3E-10	9.50E-06	
Crater Flat Group									
Prow Pass Member									
20. Unit PP4 (47 samples)	partially welded, zeolitic	32.5 ± 5	30.8 ± 5			1.104E-17	9.6E-11 ± 1.4E-10	8.29E-06	
21. Unit PP3 (166 samples)	partially welded	30.3 ± 4	16.5 ± 9			3.335E-17	2.9E-10 ± 4.0E-10	2.51E-05	
22. Unit PP2 (140 samples)	partially welded	26.3 ± 7	24.8 ± 8			6.44E-18	5.6E-11 ± 1.1E-10	4.84E-06	
23. Unit PP1 (245 samples)	partially welded, zeolitic	28.0 ± 5	26.9 ± 5			3.565E-18	3.1E-11 ± 5.6E-11	2.68E-06	
Bullfrog Member									
24. Unit BF3 (86 samples)	partially welded	11.5 ± 4	11.2 ± 4			2.415E-19	2.1E-12 ± 2.4E-12	1.81E-07	
25. Unit BF2 (65 samples)	nonwelded	25.9 ± 8	26.1 ± 9			5.75E-18	5.0E-11 ± 7.0E-10	4.32E-06	

(continued)

TABLE 4.1.1. (continued)

Locations/Units Sampled	Degree of welding	Matrix porosity (%)	Water content (%)	Dry bulk density (g/cm³)	Matrix permeability (reported units)	Matrix permeability (m²)	Hydraulic conductivity (m/sec)	Hydraulic conductivity (m/day)	Source
G. Yucca Mountain Tuffs (continued)									
Bedded Tuffs									
26. Unit BT4 (33 samples)	nonwelded	43.9 ± 12	21.6 ± 7			4.715E-14	4.1E-07 ± 1.8E-05	3.54E-02	
27. Unit BT3 (85 samples)	nonwelded	41.1 ± 8	21.6 ± 7			8.97E-14	7.8E-07 ± 1.1E-06	6.74E-02	
28. Unit BT2 (171 samples)	nonwelded	48.9 ± 11	18.5 ± 7			1.955E-13	1.7E-06 ± 2.9E-06	1.47E-01	
29. Unit BT1a (36 samples)	nonwelded	28.8 ± 7	26.7 ± 8			6.9E-18	6.0E-11 ± 8.0E-10	5.18E-06	
30. Unit BT1 (43 samples)	nonwelded	27.3 ± 7	8.3 ± 2			7.015E-15	6.1E-08 ± 3.2E-07	5.27E-03	
31. Unit BT (69 samples)	nonwelded	26.6 ± 4	26.5 ± 4			1.61E-18	1.4E-11 ± 2.1E-11	1.21E-06	Foster et al. (1985)
H. Valle Central, Costa Rica, C. A.									
andesitic tuffs (values estimated from plot)									
unweathered ignimbrite		~10						~2.5E-5	
unweathered ignimbrite		~13						~8.0E-5	
unweathered ignimbrite		~14						~8.5E-5	
unweathered ignimbrite		~18						~8.8E-5	
unweathered ignimbrite		~22						~2.0E-4	
unweathered ignimbrite		~28						~4.0E-3	
unweathered ignimbrite		~29						~7.0E-3	
unweathered ignimbrite								~8.0E-3	
unweathered ignimbrite								~6.0E-2	
I. Trans Pecos Texas									
Santana Tuff	partially welded	18–28			Avg. = 55.3 (md)	5.51E-14		4.62E-02	Fuller and Sharp (1992)
J. Chihuahua, Mexico									
Nopal Tuff (hydrothermally altered)		25.5		1.847		1.09E-16	9.49E-10	8.20E-05	Greene et al. (1995)
(mean values from 5 tuff samples)		21.0		2.034		4.70E-17	4.09E-10	3.53E-05	
		18.3		2.050		2.24E-17	1.95E-10	1.68E-05	
		8.3		2.287		1.25E-19	1.09E-12	9.42E-08	
		7.8		2.374		2.50E-19	2.17E-12	1.87E-07	
K. New Mexico					(cm³)				
Tshirege Member of Bandelier Tuff									
core sample depth = 4.2 m	variably welded				5.35E-07	5.35E-11		4.53E+01	Kearl et al. (1990)
core sample depth = 5.2 m	variably welded				6.20E-09	6.20E-13		5.25E-01	
core sample depth = 15.8 m	variably welded				2.70E-09	2.70E-13		2.29E-01	
core sample depth = 23.2 m	variably welded				1.30E-09	1.30E-13		1.10E-01	
core sample depth = 30.8 m	variably welded				1.30E-08	1.30E-12		1.10E+00	
core sample depth = 34.0 m	variably welded				1.80E-09	1.80E-13		1.52E-01	
L. Yucca Mountain, Test Well UE-25b#1									
core from testhole									
Paintbrush Group									
Topopah Springs Tuff									
1. depth = 225.7 m	densely welded	11.8	2.9	2.25		1.10E-18		8.3E-07 (K_h) / 1.6E-06 (K_v)	Lahoud et al. (1984)

(continued)

TABLE 4.1.1. (*continued*)

Locations/Units Sampled	Degree of welding	Matrix porosity (%)	Water content (%)	Dry bulk density (g/cm³)	Matrix permeability (reported units)	Matrix permeability (m²)	Hydraulic conductivity (m/sec)	Hydraulic conductivity (m/day)	Source
L. Yucca Mountain, Test Well UE-25b#1 (*continued*)									
Calico Hills Formation									
2. depth = 479.3 m	nonwelded and zeolitized	25.2	11.9	1.75		2.26E-16		**1.7E-04 (K_h)** **3.7E-05 (K_v)**	
Crater Flat Group									
Prow Pass Member									
3. depth = 625.8 m	nonwelded	23.8	10.8	1.73		8.78E-17		**6.6E-05 (K_h)** **1.8E-05 (K_v)**	
4. depth = 679.5 m	partially welded and zeolitized	10.1	3.8	2.09		1.10E-18		**8.3E-07 (K_h)** **1.7E-06 (K_v)**	
Bullfrog Member									
5. depth = 752.7 m	partially welded	23.0	9.8	2.01		8.78E-16		**6.6E-04 (K_h)** **5.0E-04 (K_v)**	
6. depth = 788.9 m	partially welded	21.3	8.8	2.06		8.78E-17		**6.6E-05 (K_h)** **7.3E-05 (K_v)**	
7. depth = 814.1 m	partially welded	22.3	8.2	2.04		6.65E-17		**5.0E-05 (K_h)** **5.0E-05 (K_v)**	
8. depth = 843.5 m	partially welded	23.1	9.1	2.02		1.10E-17		**8.3E-06 (K_h)** **1.7E-06 (K_v)**	
Tram Member									
9. depth = 923.9 m	partially welded	19.1	8.2	2.11		3.72E-17		**2.8E-05 (K_h)** **3.3E-05 (K_v)**	
10. depth = 948.8 m	partially welded	18.9	7.3	2.14		2.39E-17		**1.8E-05 (K_h)** **2.0E-05 (K_v)**	
11. depth = 1141.4 m	partially welded and zeolitized	14.7	5.1	2.24		3.86E-16		**2.9E-04 (K_h)** **3.3E-04 (K_v)**	
12. depth = 1171.0 m	partially welded and zeolitized	12.5	4.9	2.32		3.72E-16		**2.8E-04 (K_h)** **1.7E-04 (K_v)**	
Bedded Tuff									
13. depth = 1201.8 m	zeolitized	10.9	3.5	2.42		1.46E-16		**1.1E-04 (K_h)** **4.7E-05 (K_v)**	
M. Yucca Mountain									
1. Tiva Canyon Tuff (1 sample)	dense	8				**1E-18**		8.47E-07	Moyer et al. (1996)
2. Tiva Canyon (1)	moderate	20				**5E-17**		4.23E-05	
3. Tiva Canyon (1)	non to partial	40				**1E-13**		8.47E-02	
4. Pre-Tiva Canyon Bedded Tuff(1)	nonwelded	44				**1E-13**		8.47E-02	
6. Yucca Mountain Tuff (1)	nonwelded	42				**5E-14**		4.23E-02	
7. Yucca Mountain Tuff (1)	moderate	20				**5E-15**		4.23E-03	

(*continued*)

TABLE 4.1.1. (continued)

Locations/Units Sampled	Degree of welding	Matrix porosity (%)	Water content (%)	Dry bulk density (g/cm³)	Matrix permeability (reported units)	Matrix permeability (m²)	Hydraulic conductivity (m/sec)	Hydraulic conductivity (m/day)	Source
M. Yucca Mountain (*continued*)									
8. Yucca Mountain Tuff (1)	nonwelded	32				1E-13		8.47E-02	
9. Pre-Yucca Mountain Tuff, bedded tuff (1)	nonwelded	38				1E-13		8.47E-02	
10. Pah Canyon Tuff (1)	moderate to non	48				1E-13		8.47E-02	
11. Pre-Pah Canyon Tuff, bedded tuff (1)	nonwelded	55				5E-12		4.23E+00	
12. Topopah Spring Tuff, pumice fall (1)	non to partial	55				1E-12		8.47E-01	
13. Topopah Spring Tuff, pumice fall (1)	moderate	28				1E-15		8.47E-04	
14. Topopah Spring Tuff, pyroclastic flow (1)	dense	3				1E-19		8.47E-08	
15. Topopah Spring Tuff, pyroclastic flow (1)	dense	8				1E-18		8.47E-07	
16. Topopah Spring Tuff, pyroclastic flow (1)	dense	18				1E-16		8.47E-05	Peters et al. (1984)
N. Yucca Mountain Tuff									
1. 1 sample	vitric, nonwelded	**40**				**4.49E-14**	**3.90E-07**	3.37E-02	
2. 1 sample	vitric, nonwelded	**46**				**3.11E-16**	**2.70E-09**	2.33E-04	
3. 1 sample	dense	**8**				1.12E-18	9.70E-12	8.38E-07	
4. 1 sample	dense	**11**				2.19E-18	1.90E-11	1.64E-06	
5. 1 sample	dense	**7**				1.73E-19	1.50E-12	1.30E-07	
6. 1 sample	zeolitized, nonwelded	28				2.30E-18	2.00E-11	1.73E-06	Rasmussen et al. (1993)
O. Arizona, White Unit of Apache Leap Tuff (mean of 150 samples)		17.5		2.1					
1. matrix water permeability						**5.60E-16**		4.74E-04	
2. matrix air permeability						**1.77E-15**		1.50E-03	
P. Yucca Mountain Tiva Canyon Tuff							**log(K) (m/sec)**		
1. Upper Ash-Flow Unit (mean of 122–130)	dense	19.5		1.89		2.15E-14	-6.728	1.62E-02	
2. Lower Ash-Flow Sub-Unit (mean of 133–140)	nonwelded	40.3		1.38		1.71E-11	-3.826	1.29E+01	
3. Basal Air-Fall Tuff Sub-Unit (mean of 33–36)	nonwelded	52		1.13		1.61E-10	-2.862	1.21E+02	Rautman et al. (1995)

(continued)

TABLE 4.1.1. (continued)

Locations/Units Sampled	Degree of welding	Matrix porosity (%)	Water content (%)	Dry bulk density (g/cm³)	Matrix permeability (reported units)	Matrix permeability (m²)	Hydraulic conductivity (m/sec)	Hydraulic conductivity (m/day)	Source
Q. New Mexico									Rodgers and Gallaher (1995)
Tshirege Member of Bandelier Tuff									
unit 3 (vapor phase alteration)	non to moderate	46.9		1.47		1.01E-13	8.80E-07	7.60E-02	
unit 2b/2 (vapor phase alteration)	moderate to dense	47.9		1.37		7.48E-13	6.50E-06	5.62E-01	
unit 2a/1v-u (pumiceous, vap. phase alt.)	non to partial	48.2		1.27		2.19E-13	1.90E-06	1.64E-01	
unit 1b/1v-c (pumiceous, vap. phase alt.)	non to partial	52.8		1.18		1.96E-13	1.70E-06	1.47E-01	
unit 1a/1g (pumiceous tuff, glassy)	non to partial	46		1.26		2.65E-13	2.30E-06	1.99E-01	
Tsankawi pumice bed and Cerro Toledo		49		1.25		1.50E-12	1.30E-05	1.12E+00	
Otowi Member (ignimbrite)	nonwelded	46.9		1.18		7.25E-13	6.30E-06	5.44E-01	
Bandelier Tuff (all)		49.2		1.22		4.49E-13	3.90E-06	3.37E-01	
R. Yucca Mountain Tuff Outcrops				grain density (g/cm³)					Scott et al. (1983)
Paintbrush Group									
Tiva Canyon Tuff									
(13 samples)—most mafic	partially welded	19.7 ± 6		2.51 ± 0					
(33 samples)—most devitrified	partially welded	9.8 ± 3		2.49 ± 0					
(10 samples)—partly devitrified	partially welded	9.6 ± 5		2.45 ± 0					
(33 samples)—least devitrified	partially welded	38.0 ± 13		2.38 ± 0					
S. Trans-Pecos Texas					(md)				Smyth-Boulton (1995)
1. Santana Tuff									
(Arithmetic mean 8 poro./87 perm. samples)	nonwelded	41		1.4	929.1	9.17E-13		7.76E-01	
(Range)		38–42		0.94–1.52	579–1360	5.71E-13 to 1.34E-12		4.84E-01 to 1.14E+00	
2. Santana Tuff									
(Geom. mean of 25 poro./40 perm. samples)	partially welded	20		1.9	16.3	1.61E-14		1.36E-02	
(Range)		12–28		1.81–2.12	0.95–97.06	9.38E-16 to 9.58E-14		7.94E-04 to 8.11E-02	

(continued)

TABLE 4.1.1. (continued)

Locations/Units Sampled	Degree of welding	Matrix porosity (%)	Water content (%)	Dry bulk density (g/cm³)	Matrix permeability (reported units)	Matrix permeability (m²)	Hydraulic conductivity (m/sec)	Hydraulic conductivity (m/day)	Source
S. Trans-Pecos Texas (*continued*)									
<u>3. Santana Tuff</u>									
(Geom. mean of 18 poro./ 24 perm samples)	densely welded	7.5		2.3	0.1	9.87E-17		8.36E-05	
(Range)		4–12		2.14–2.41	0.01–1.99	9.87E-18 to 1.96E-15		8.36E-06 to 1.66E-03	
T. Yucca Mountain, Test Well USW H-1 core from testhole									Weeks and Wilson (1984)
<u>Paintbrush Group</u> depth = 33.5 m	nonwelded	42.8	22.0	1.38					
<u>Topopah Spring Tuff</u> (16 samples)	densely welded	15.5 ± 3	10.4 ± 1	2.19 ± 0.8					
depth = 128.0 m to 405.8 m									
<u>Calico Hills Formation</u> depth = 530.7 m	nonwelded	46.1	45.9	1.30					
depth = 532.8 m	nonwelded	40.9	42.2	1.43					

TABLE 4.1.2.

Locations / Units Sampled	Degree of welding	Permeability (cm²)	Arithmetic mean permeability (m²)	Transmissivity (m²/day)	Thickness of tested interval (m)	Hydraulic conductivity (m/day)	Source
A. Yucca Mountain, Test Well UE-25p#1							Craig and Robison (1984)
aquifer pumping test results using intervals determined from borehole flow surveys							
Prow Pass Member of Crater Flat Group	highly fractured			14	30	4.7E-01	
packer-injection test results for entire section of saturated Tertiary volcanics.							
depth interval tested: 380 m–1180 m	variably welded			>30	800	>3.8E-02	
aquifer pumping test results for fracture permeability of entire saturated section of Tertiary volcanics.							
minimum value	variably welded			18	919	2.0E-02	
maximum value	variably welded			60	919	6.5E-02	
most likely representative value	variably welded			25	919	2.7E-02	
B. Yucca Mountain, Test Well USW H-6							Craig and Reed (1991)
aquifer pumping test results using intervals determined from borehole flow surveys							
Middle Bullfrog Member of Crater Flat Gp. (60% of total well production)	partially welded			140	15	9.3E+00	
Middle Tram Member of Crater Flat Gp. (32% of total well production)	partially welded			75	11	6.8E+00	
aquifer pumping test results for entire section of saturated Tertiary volcanics.							
pump test 1 (range) 526-1220 m	nonwelded to densely welded			240 480	694 694	3.5E-01 6.9E-01	
pump test 2	nonwelded to densely welded			230	694	3.3E-01	
pump test 3 (interval restricted by packers) Tram Member of Crater Flat Tuff (753–834 m)	nonwelded to partially welded			76	81	9.4E-01	
C. Bandung Basin, Indonesia							Directorate of Environmental Geology, Department of Mine and Energy, Indonesia
Tangkuban Prahu Volcanics (Tuff, Lava, and Sand)				50–900			
Southern Volcanics (Volcanic Breccia and Lahars)				2–100			Data provided via personal communication from Eddie Listancia (1995)
Eastern Volcanics of the Mandalawangi (Volcanic Breccia and Tuffaceous Sand)				40–800			

(continued)

TABLE 4.1.2. (*continued*)

Locations / Units Sampled	Degree of welding	Permeability (cm²)	Arithmetic mean permeability (m²)	Transmissivity (m²/day)	Thickness of tested interval (m)	Hydraulic conductivity (m/day)	Source
D. Tshirege Member of Bandelier Tuff							Kearl et al. (1990)
depth to test interval in well = 3.3 m	variably welded	1.1E-08				9.3E-01	
depth to test interval in well = 7.9 m	variably welded	1.4E-08				1.2E+00	
depth to test interval in well = 11.3 m	variably welded	1.5E-08				1.3E+00	
depth to test interval in well = 22.2 m	variably welded	1.7E-08				1.4E+00	
depth to test interval in well = 12.5 m	variably welded	8.4E-09				7.1E-01	
depth to test interval in well = 29.3 m	variably welded	1.0E-08				8.5E-01	
depth to test interval in well = 27.1 m	variably welded	1.6E-08				1.4E+00	
depth to test interval in well = 25.8 m	variably welded	6.5E-08				5.5E+00	
depth to test interval in well = 24.1 m	variably welded	2.9E-08				2.5E+00	
depth to test interval in well = 18.9 m	variably welded	8.0E-08				6.8E+00	
depth to test interval in well = 4.2 m	variably welded	1.9E-07				1.6E+01	
depth to test interval in well = 15.8 m	variably welded	4.3E-08				3.6E+00	
depth to test interval in well = 23.2 m	variably welded	6.5E-09				5.5E-01	
depth to test interval in well = 34.0 m	variably welded	1.1E-08				9.3E-01	
E. Yucca Mountain, Test Well UE-25b#1							Lahoud et al. (1984)
packer-injection (slug) test results							
Calico Hills Tuff (3 different intervals)	non to partially welded zeolitized			6.9 1.7 1.4	14 14 40	4.9E-01 1.2E-01 3.5E-02	
Prow Pass Member of Crater Flat Group	non to partially welded devitrified, faulted			2.1	40	5.3E-02	
Lower Prow Pass and Upper Bullfrog Members of Crater Flat Group	bedded air-fall partially welded, devitrified			0.18	40	4.5E-03	
Bullfrog Member of Crater Flat Group	partially welded, devitrified			0.099	40	2.5E-03	
Lower Tram Member of Crater Flat Group and Lithic Ridge Tuff	partially welded, bedded moderately welded			0.021	214	9.8E-05	
aquifer pumping test results							
Entire Saturated Interval (1 pump test)	variably welded			340	749	4.5E-01	

(*continued*)

TABLE 4.1.2. (continued)

Locations / Units Sampled	Degree of welding	Permeability (cm²)	Arithmetic mean permeability (m²)	Transmissivity (m²/day)	Thickness of tested interval (m)	Hydraulic conductivity (m/day)	Source
F. Yucca Mountain Boreholes air-injection testing							LeCain (1997)
Paintbrush Group							
Tiva Canyon Tuff							
Borehole UZ-16 (4 intervals)	partially welded		1.2E-11			1.0E+01	
Borehole SD-12 (11 intervals)	partially welded		7.0E-12			5.9E+00	
Borehole NRG-6 (4 intervals)	partially welded		1.1E-11			9.5E+00	
Borehole NRG-7a (4 intervals)	partially welded		2.7E-11			2.3E+01	
Borehole NRG-7a (2 intervals)	nonwelded		2.0E-13			1.7E-01	
Yucca Mountain Tuff							
Borehole NRG-7a (4 intervals)	nonwelded		3.00E-13			2.5E-01	
Pah Canyon Tuff							
Borehole NRG-7a (7 intervals)	nonwelded		2.00E-13			1.7E-01	
Bedded Tuffs							
Borehole NRG-7a (1 interval)	nonwelded		2.0E-13			1.7E-01	
Borehole NRG-7a (1 interval)	nonwelded		3.0E-12			2.5E+00	
Borehole NRG-7a (1 interval)	nonwelded		7.0E-13			5.9E-01	
Topopah Spring Tuff							
Borehole UZ-16 (54 intervals)	densely welded		1.8E-12			1.5E+00	
Borehole SD-12 (27 intervals)	densely welded		4.7E-12			4.0E+00	
Borehole NRG-6 (34 intervals)	densely welded		2.1E-12			1.8E+00	
Borehole NRG-7a (38 intervals)	densely welded		4.0E-13			3.4E-01	
G. Yucca Mountain, Test Well USW G-4							Lobmeyer (1986)
packer-injection (slug) test results							
sum of packer-injection (slug) tests	variably welded			7	234	3.0E-02	
Prow Pass Member of Crater Flat Group	non to partially welded devitrified and zeolitic			1.16	39	3.0E-02	
Bullfrog Member of Crater Flat Group (2 intervals)	partially welded, devitrified with shear fractures			0.51 / 0.63	46 / 25	1.1E-02 / 2.5E-02	
Bullfrog Member of Crater Flat Group (2 intervals)	non to partially welded, devitrified, ash-fall bedded and altered			1.15	45	2.6E-02	
lower Bullfrog and upper Tram Members, Crater Flat Group	non to partially welded, devitrified and bedded			0.76	46	1.7E-02	
				2.3	24	9.6E-02	

(continued)

TABLE 4.1.2. (continued)

Locations / Units Sampled	Degree of welding	Permeability (cm²)	Arithmetic mean permeability (m²)	Transmissivity (m²/day)	Thickness of tested interval (m)	Hydraulic conductivity (m/day)	Source
G. Yucca Mountain, Test Well USW G-4 (*continued*)							
aquifer pumping and recovery test results							
Crater Flat Group (entire saturated interval)	variably welded						
pumping test				622	518	1.20E+00	
recovery test				490	377	1.30E+00	
pumping test				675	482	1.40E+00	
recovery test				570	380	1.50E+00	
reported average for test well				600	374	1.60E+00	
aquifer pumping test results using intervals determined from borehole flow surveys							
98% of average flow coming from fractured interval in Tram Member, Crater Flat Group	partially welded, devitrified and zeolitized			588	65	9	
75% of average flow coming from fractured interval in the Bullfrog and upper Tram members of the Crater Flat Group	devitrified with slicken-sided fractures			450	10	45	
							Thordarson et. al. (1985)
H. Yucca Mountain, Test Well USW H-3							
results from 3 aquifer pumping tests Tram Member of Crater Flat Group and upper Lithic Ridge Tuff	partially welded, fractured and devitrified			0.5	465	1.0E-03	
				0.4	465	9.0E-04	
				1.0	397	2.5E-03	
							Whitfield et al. (1985)
I. Yucca Mountain, Test Well USW H-4							
aquifer pumping test results using intervals determined from borehole flow surveys							
Prow Pass Member of Crater Flat Group (4 different intervals)	partially welded			11.5	24	4.9E-01	
	partially welded			18.6	16	1.2E+00	
	partially welded			6.9	21	3.3E-01	
	bedded, zeolitized			1.1	27	4.0E-02	
Bullfrog Member of Crater Flat Group (3 different intervals)	highly fractured zone			55.9	26	2.2E+00	
	non to partially welded			8.2	6	1.4E+00	
	non to partially welded			10.4	20	5.2E-01	
Tram Member of Crater Flat Group (5 different intervals)	non to partially welded			5.7	14	4.1E-01	
	non to partially welded			3.4	18	1.9E-01	
	highly fractured zone			53.5	53	1.0E+00	
	non to partially welded			1.2	6	2.0E-01	
	non to partially welded			1.4	46	3.0E-02	
Lithic Ridge Tuff (4 different intervals)	partially welded and zeolitized			20.4	13	1.6E+00	
				1.3	19	7.0E-02	
				1.6	6	2.7E-01	
Entire saturated thickness of test well (range of values)	variably welded			200	680	2.9E-01	
				790	680	1.2E+00	

however, the most complete measurements of fractures in tuffs are from outcrop studies presented elsewhere in this book (see Chapter 2.3, this volume). The significance of fracture flow to the transmission of fluids in tuffs becomes apparent through comparison of data from matrix versus well test methods.

Values appearing in bold type in both tables are those reported in the references. Regardless of whether values were reported in units of permeability, hydraulic conductivity, or transmissivity, we have converted all values to equivalent saturated hydraulic conductivity units of meters per day (m/d) for ease of comparison between different researchers and different measurement methodologies. Values converted to units other than those reported are in plain type. We tabulate the data alphabetically by author.

Laboratory Measurements

Hydrologic parameters included in our table of tuff matrix data (Table 4.1.1) are: (1) porosity, (2) water content, (3) bulk density, (4) permeability, and (5) saturated hydraulic conductivity. The most common methods that have been used by researchers to measure porosity are gravimetric, helium pycnometry, and mercury-injection porosimetry. Permeability and saturated hydraulic conductivity have been measured using air-permeameters and a variety of constant head permeameter apparatus, and estimated using inverse numerical modeling. In the first column of Table 4.1.1, we indicate the geologic unit and whether the hydrologic values represent a single sample or an average value of multiple samples. In the second column, we give the degree of welding reported by researchers, even though it appears that this parameter may have been ambiguously determined.

Researchers responsible for characterizing the hydrologic properties of Yucca Mountain started conducting laboratory analyses of tuff matrix samples in the early 1980s. Montazer and Wilson (1984) were typical of researchers in the 1980s, who apparently assumed a degree of homogeneity when they divided the 500–750-ft-thick sequence of unsaturated zone tuffs at Yucca Mountain into five hydrologic units. The five units, Tiva Canyon welded, Paintbrush Canyon nonwelded, Topopah Spring welded, Calico Hills nonwelded, and Crater Flat tuffs, were separated in part on the basis of degree of welding and number of fractures per unit volume of rock. Other early workers (i.e., Winograd, 1971) also recognized that nonwelded tuffs have higher porosity and permeability than densely welded tuffs, and that fractures in densely welded tuffs allow sufficient storage and transmission of water for these units to be designated as aquifers. They designated zones of nonwelded tuff as aquitards by assuming that capillary or suction forces within these zones of relatively low water content would prevent transmission of water either through the matrix or along fractures. As a result of early assumptions, most of the laboratory analyses were performed on either nonwelded or densely welded tuff samples (i.e., Dailey et al., 1987; Weeks and Wilson, 1984; Table 4.1.1). Researchers began recognizing how tuff heterogeneities could affect fluid flow and further categorized nonwelded tuffs as vitric nonwelded or nonwelded

zeolitized for laboratory analyses of core samples (Lahoud et al., 1984; Peters et al., 1984; Table 4.1.1). Lahoud et al. (1984) also described and analyzed samples of partially welded and devitrified Yucca Mountain tuffs. In more recent studies, researchers have further subdivided the Yucca Mountain tuffs into 15 lithostratigraphic units (i.e., Moyer et al., 1996; Table 4.1.1) and 31 hydrologic units (Flint, 1998; Table 4.1.1) to more accurately represent the heterogeneity that needs to be considered in any realistic numerical model of groundwater flow.

Even though degree of welding has been a major criterion for subdividing tuff hydrologic units at Yucca Mountain, it has not been consistently defined. For example, Rautman et al. (1995) presented data for densely welded samples of Tiva Canyon Tuff with porosity of 19.5% and permeability of 2.15×10^{-14} m^2 (sample P-1 on Table 4.1.1). Whereas Moyer et al. (1996) report values of 8% porosity and 1.0×10^{-18} m^2 permeability for the same densely welded unit (sample M-1 on Table 4.1.1). Welding of tuffs has primarily been defined qualitatively on the basis of (1) color, (2) presence or absence and density of jointing, (3) stratigraphic relationship to adjacent layers, and (4) lithification (i.e., Moyer et al., 1996). Welding can be quantified through laboratory measurement of porosity and bulk density (Smith and Bailey, 1966; Wohletz and Heiken, 1992) or by estimating flattening (increased length-to-width ratio) of pumice fragments in ash-flow tuffs (Peterson, 1979). Smyth-Boulton (1995) used the following laboratory methods to quantify degree of welding in the Santana Tuff of Trans-Pecos Texas: (1) mercury porosimetry for porosity and bulk density and (2) petrographic analysis for measuring length-to-width ratios of pumice. Santana Tuff samples were first categorized as nonwelded, partially welded, or densely welded (NDP) on the basis of pumice flattening. Probability plots of mercury-porosimetry and air-permeability measurements for the NPD-welded samples suggest three populations of log-normally distributed data (Smyth-Boulton, 1995) (Table 4.1.1). Flint (1998) used porosity measurements to help subdivide the Yucca Mountain tuffs into 31 distinct hydrologic units.

Studies of petroleum and geothermal reservoirs show that effective porosity and intrinsic permeability usually correlate. Deviations from this correlation are thought to be due to differences in pore geometry, pore-filling cements, and fracture permeability (Björnsson and Bodvarsson, 1990; Shenhav, 1971). Most researchers represented in Table 4.1.1 have reported both porosity and permeability data for the same sample or set of samples, so it is possible to evaluate the porosity-permeability relationships of tuffs. Our plot of permeability in units of meters squared versus percent porosity includes data from 10 of the total 20 referenced data sets (Fig. 4.1.1).

There is good correlation between permeability and porosity of the tuff data, especially considering that the tuff permeabilities range over 11 orders of magnitude. Low porosity deviations from this correlation come from samples for which the pore spaces are most likely not well connected. Sample N-2 (Table 4.1.1, Fig. 4.1.1) from the Peters et al. (1984) reference is vitric and nonwelded, with porosity of 46% and permeability

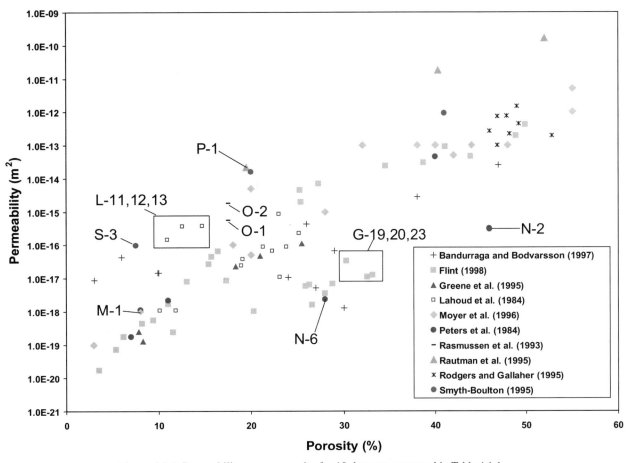

Figure 4.1.1. Permeability versus porosity for 10 data sets presented in Table 4.1.1.

of 3.11×10^{-16} m^2. Sample N-6 is described as nonwelded and zeolitized, with porosity of 28% and permeability of 2.3×10^{-18} m^2 (Table 4.1.1, Fig. 4.1.1). These samples, and others that fall within the higher porosity–lower permeability range (i.e., G-19, G-20, and G-23 on Table 4.1.1, Fig. 4.1.1), must have more isolated pore spaces that do not allow transmission of fluids. Studies of fluid flow through rocks of differing porosities show that samples with larger, more equidimensional pores give more consistent permeability measurements than samples that have more tortuous pore networks, because fluids flow through the center and are not deflected by pore-edge effects. According to Pettijohn et al. (1987), both length of flow path through a sediment (i.e., tortuosity) and roughness of pore walls contribute to a decrease in permeability relative to effective porosity. The presence of zeolites or devitrification mineralization in tuffs would increase pore surface roughness and tortuosity and, thereby, cause a decrease in intrinsic permeability of the rock. There are some zeolitized samples with higher permeability relative to porosity (L-11, L-12, and L-13 in Table 4.1.1, Fig. 4.1.1); however, Lahoud et al. (1984) reported that they came from an interval that was identified by borehole flow testing (methodology discussed in next section) as being highly fractured. Presence of microfractures in

the more densely welded samples (those in the lower porosity range) also results in higher permeability relative to porosity (i.e., sample S-3). Smyth-Boulton (1995) examined thin sections of densely welded Santana Tuff that had been vacuum-impregnated with blue-dyed epoxy and observed micrometer-scale fracturing in the densely welded and brittle rocks.

Another factor that could cause deviation from the permeability-porosity correlation is the permeability measurement method. The two measurements provided by Rasmussen et al. (1993) are both from the same sample of partially welded Apache Leap Tuff. The value determined using an air-permeameter is 1.77 $\times 10^{-13}$ m^2 (O-2) versus the result of 5.6×10^{-16} m^2 (O-1) using a water-based permeameter. Klinkenberg gas slippage effects (Klinkenberg, 1941) may be the reason why some of the samples measured using air-permeability apparatus have relatively higher permeabilities (i.e., Kearl et al. [1990] in Table 4.1.1).

In Situ Measurements

Tuff hydrologic data in Table 4.1.2 are primarily results of in situ testing conducted in the 1980s by United States Geological Survey (USGS) hydrologists who supervised drilling and testing of six deep wells (915 m to 1805 m) to determine saturated zone hydrologic properties of Yucca Mountain tuffs. Designations for wells used in these early tests are UE-25p#1, USW H-6, UE-25b#1, USW G-4, USW H-3, and USW H-4. Data presented in Table 4.1.2 are listed alphabetically by author. The in situ testing methods used by USGS researchers included (1) geophysical logging, (2) aquifer pumping and recovery tests, (3) radioactive tracer and temperature borehole flow surveys, and (4) packer-injection (slug) tests. Detailed descriptions of aquifer pumping and recovery testing and analysis are given in Kruseman and de Ridder (1990). A good description of packer-injection testing can be found in Driscoll (1986). More recently, researchers have used in situ air-injection for unsaturated zone hydraulic testing in shallow boreholes drilled in the Bandelier Tuff (Kearl et al., 1990) and Yucca Mountain tuffs (LeCain, 1997).

Hydrologic data from well tests are valuable because they give a macroscopic view of groundwater flow through both rock matrix and fractures. The USGS hydrologists used geophysical logging of boreholes prior to installation of casing as a way to determine relative porosity of subsurface units. They used these results as another way to select intervals to refine estimates of hydrologic properties in the test wells. Results from aquifer pumping and recovery tests are primarily presented as transmissivity, which is a measure of groundwater flow through a specified interval (equations 4.8 and 4.9 herein). Several authors (e.g., Whitfield, et al., 1985) also gave thickness of tested interval and estimated hydraulic conductivities. Since total saturated intervals tested by the USGS are very large, results do not provide hydrologic properties of individual tuff units. For example, the range of total saturated intervals tested in the USGS wells is 465 m in USW H-3 (Thordarson et al., 1985) to 919 m in UE-25p#1 (Craig and Robison, 1984). These intervals span variably welded subunits of at least four distinct tuff units. This scale of hydrologic information is useful from the standpoint of water supply but may not be detailed enough for assessing potential groundwater contamination issues, and the method is limited to the saturated zone.

Results from in situ methods using radioactive tracer and temperature borehole flow surveys and packer-injection (slug) tests can be used to refine knowledge of hydrologic properties of more specific tuff intervals, especially when combined with descriptions of cores extracted from the test holes during drilling. Radioactive tracer borehole surveys are described in detail by Craig and Robison (1984). The USGS hydrologists used radioactive tracer borehole flow and borehole temperature instrumentation to identify higher flow-rate intervals within the test wells. They then partitioned aquifer pumping test results over these discrete intervals to obtain more refined estimates of hydrologic properties of particular tuff units or zones within units. For example, Lobmeyer (1986) reported an average transmissivity of 600 m^2/d for a 374-m-thick interval of Crater Flat Tuff in test well USW G-4; the hydraulic conductivity for this interval is 1.6 m/d (Table 4.1.2). Borehole flow surveys showed that 75% of total flow in the test well USW G-4 was coming from a 10-m-thick interval in the Tram Member of the Crater Flat Group, which

was described as devitrified with slickensided fractures. Assigning 75% of the average transmissivity in the well to this 10 m interval gave a value of 450 m^2/d and a much higher hydraulic conductivity of 45 m/d. Other USGS hydrologists used this same methodology to refine their estimates of hydrologic properties for individual tuff units. Transmissivity and hydraulic conductivity values determined using this methodology are tabulated under the subheading "Aquifer pumping test results using intervals determined from borehole flow surveys" in Table 4.1.2.

Packer-injection or slug testing was another way the USGS hydrologists estimated hydraulic properties of individual tuff units. They used results of borehole flow surveys to define intervals of hydrologic interest and then placed packers (inflatable seals) at specified depths in the wells. A "slug" of water introduced between the packers caused the water level in that interval to rise nearly instantaneously, and the rate of water-level decline was monitored to determine the ability of the tuff units adjacent to the specific interval to soak up the introduced water, thereby allowing estimation of the interval's hydraulic conductivity. According to Craig and Robison (1984), leaking packers and limitations of the monitoring tools at great depths within the wells made results from the packer-injection tests suspect. Because of this uncertainty, we do not report their transmissivity values determined using this method. Lahoud et al. (1984) and Lobmeyer (1986) more confidently reported results obtained using the packer-injection methodology, so we do include their values in Table 4.1.2.

Lahoud et al. (1984) calculated a hydraulic conductivity of 4.5×10^{-1} m/d for the entire saturated thickness in well UE-25b#1. This value is nearly equal to and one to four orders of magnitude higher than the packer-injection–derived hydraulic conductivities reported for individual intervals (e.g., 3.5×10^{-2} to 4.9×10^{-1} m/d for isolated intervals within the Calico Hills Tuff, and 9.8×10^{-5} m/d for the 214-m-thick interval spanning the lower Tram Member of Crater Flat Group and Lithic Ridge Tuff). Packer-injection test results are likely more representative of tuff matrix and localized brittle deformation, whereas the aquifer pumping test results represent combined matrix and much larger-scale fracture hydrologic properties. The same argument can be made for the difference between packer-injection and entire-well hydraulic conductivity values reported by Lobmeyer (1986). In USW G-4, hydraulic conductivities estimated from packer-injection tests of 24–46-m-thick isolated intervals are two orders of magnitude lower than the reported average hydraulic conductivity of 1.6 m/d for a 374-m-thick, saturated interval (Table 4.1.2).

Total well hydraulic conductivities of thick sequences of Yucca Mountain tuffs of the Crater Flat Group downward into the Lithic Ridge Tuff and older tuffs range from 10^{-4} to 10^{1} m/d. Some of this variation most likely is a result of real differences in the substrate (i.e., heterogeneity of the tuff units), but some of it must be a result of testing and analysis methods. For example, the very low hydraulic conductivities estimated for well USW H-3 could be from poor well performance. Significant portions of this well are reported to have collapsed during drilling (Thordarson,

1985). It is very difficult to completely clean out a well after collapse occurs, especially in a well that is over a kilometer deep. If we ignore the results from test well USW H-3, the range of hydraulic conductivity determined using aquifer pumping tests of the entire saturated thickness in USGS test wells is 10^{-2} to 10^{1} m/d. Air testing in boreholes at Yucca Mountain (LeCain, 1997) and in the Bandelier Tuff in New Mexico (Kearl et al., 1990) yielded higher average values of hydraulic conductivity than the aquifer pumping tests. These values ranged from 1.71×10^{-1} to 9.5 m/d even for variably welded tuffs (Table 4.1.2).

DISCUSSION

Knowledge of tuff hydrologic properties is useful for assessing groundwater resources in a variety of locations around the world (e.g., Central America, southwestern USA, and Indonesia) and for determining potential environmental impacts associated with waste disposal (e.g., Yucca Mountain, Nevada, and Los Alamos, New Mexico, USA). Hydrologic properties of tuffs vary greatly depending upon factors such as degree of welding, cooling rate, mineralogical alteration products, and postdepositional tectonic stress, but also as a result of measurement methodology. For example, laboratory measurement of matrix permeability of core samples taken from a borehole may yield values many orders of magnitude lower than the permeability calculated from an aquifer pumping test performed in the same borehole. Lahoud et al. (1984) and Peters and Klavetter (1988) recognized early in the study of hydrology of tuffs the importance of measuring hydrologic properties of both matrix samples in the laboratory and matrix plus fractures in well tests. These and other workers found that permeabilities measured in boreholes were consistently greater by several orders of magnitude than those measured in laboratories for both the Yucca Mountain tuffs (Lahoud

et al., 1984; Peters and Klavetter, 1988) and the Apache Leap Tuff (Rasmussen et al., 1993). They found that well tests not only yielded higher hydraulic conductivities, the values were more realistic because they accounted for (1) heterogeneity of the tuff matrix and (2) matrix plus fracture permeability. We have plotted the minimum and maximum values of laboratory and in situ values of tabulated hydraulic conductivity data to illustrate this point (Fig. 4.1.2).

Smyth-Boulton (1995) documented increased density of columnar cooling joints with increased degree of welding in the Santana Tuff, so fluid flow in fractures should be greater in more densely welded tuffs. Aquifer testing results show that significant groundwater movement in the saturated zone of Yucca Mountain appears to be controlled by fractured intervals. These test results show that the fluid flow is concentrated in tectonically fractured zones near faults. One perspective on this issue is a study by Pearcy et al. (1995) of uranium transport in the heavily fractured, silicic Nopal Tuff of Chihuahua, Mexico. They did not describe variations in the degree of welding in the tuff, most likely because it had been highly mineralized and altered, nor did they measure or report hydraulic properties of the tuff. However, through isotopic analyses of uranium-bearing minerals, Pearcy et al. (1995) were able to document transport distances 20 times greater along macrofractures than through a network (>1 fracture/cm) of discontinuous microfractures in the tuff. They also documented almost immeasurable (<1 mm) transport of uranium in unfractured tuff matrix perpendicular to macroscopic fractures.

Another early misconception about tuff hydrology was that nonwelded tuffs in the unsaturated zone acted as aquitards because (1) they were not very fractured and (2) the high matrix porosity and permeability and, therefore, high capillary tension of pores would inhibit water flux (Winograd, 1971). It was also thought that any fractures that were present in nonwelded tuffs

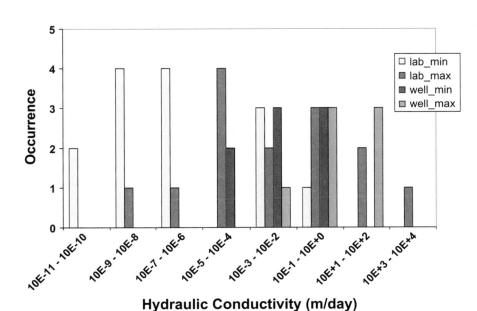

Figure 4.1.2. Ranges of hydraulic conductivity for laboratory and in situ measurements.

would not transmit water because it would be drawn into the matrix by capillary suction (termed imbibition). This assumption made sense because unsaturated zone hydraulic conductivity is dependent upon the magnitude of the negative pore pressure (moisture potential or suction pressure), which is dependent upon pore water content (θ). In other words, the ease with which water flows in the unsaturated zone increases with increasing water content because the tension under which water is held in the pores decreases with increasing water content. Early workers in hydrology at Yucca Mountain assumed that imbibition of water from fractures into matrix would prevent any significant fluid flux through the unsaturated zone. However, they did not realize that formation of skins along fracture surfaces could prevent matrix imbibition from taking place. Fracture skins are both coatings of the fracture surface by minerals, and alterations of the fracture walls created by mineral precipitation, infiltered detritus, and rock-water interactions (Sharp et al., 1995). Thoma et al. (1992) identified low permeability alteration minerals along fractures in tuffs from Yucca Mountain. Fu et al. (1994) and Sharp et al. (1995) fully characterized fracture skins in sandstone and tuff, respectively. Sharp et al. (1995) also quantified the effects of fracture skins on contaminant transport; they concluded that the ubiquitous presence of fracture skins in fractured rock systems must be considered in numerical groundwater models.

Several different aspects and scales of heterogeneity occur in tuffs. Matrix heterogeneity results from variations in (1) percentages of phenocrysts and rock fragments, (2) in degree of devitrification of glass shards and pumice, and (3) in degree of welding and compaction. Fracture heterogeneity results from (1) variations in spacing of columnar cooling joints, (2) from presence or absence of tectonic fractures, and (3) from differences in properties of fracture skins. By recognizing that the hydrologic properties of tuffs are directly related to their complex petrology, realistic data sets that incorporate heterogeneous hydrologic properties will result in more accurate groundwater models for predicting contaminant migration.

ACKNOWLEDGMENTS

Support for this work was provided in part by the John A. and Katherine G. Jackson School of Geosciences and the Geology Foundation at the University of Texas at Austin. We would also like to thank Grant Heiken for following through with the idea of publishing a book on tuffs, for serving as editor, and for his persistent yet friendly encouragement to the senior author.

REFERENCES CITED

Anderson, L.A., 1981, Rock property analysis of core samples from the Yucca Mountain UE25a-1 borehole, Nevada Test Site, Nevada: U.S. Geological Survey Open-File Report 81-1338, 31 p.

Anderson, L.A., 1994, Water permeability and related rock properties measured on core samples from the Yucca Mountain GU-3/G-3 and USW G-4 boreholes: U.S. Geological Survey Open-File Report 92-201, 36 p.

ASTM (American Society for Testing and Materials Standards), 2000, Standard test method for permeability of granular soils (constant head): ASTM D2434-68, p. 1–5.

Bandurraga, T.M., and Bodvarsson, G.S., 1997, Calibrating matrix and fracture properties using inverse modeling, *in* Bodvarsson, G.S., and Bandurraga, T.M., eds., The site scale unsaturated zone model of Yucca Mountain, Nevada, for the viability assessment: Chapter 6: Berkeley, California, Lawrence Berkeley National Laboratory, LBNL-40376, LB970601233129.001.

Björnsson, G., and Bodvarsson, G., 1990, A survey of geothermal reservoir properties: Geothermics, v. 19, p. 17–27, doi: 10.1016/0375-6505(90)90063-H.

Butler, J.J., Jr., 1998, The design, performance, and analysis of slug tests: Boca Raton, Florida, CRC Press, 252 p.

Byers, F.M., Jr., Carr, W.J., Orkild, P.P., Quinlivan, W.D., and Sargent, K.A., 1976, Volcanic suites and related cauldrons of Timber Mountain–Oasis Valley caldera complex, southern Nevada: U.S. Geological Survey Professional Paper 919, 70 p.

Chuchula, R.J., 1981, Reconnaissance geology of the Sierra Rica area, Chihuahua, Mexico [Master's thesis]: Austin, University of Texas at Austin, 167 p.

Craig, R.W., and Reed, R.L., 1991, Geohydrology of rocks penetrated by test well USW H-6, Yucca Mountain, Nye County, Nevada: U.S. Geological Survey Water Resource Investigations Report 89-4025, 40 p.

Craig, R.W., and Robison, J.H., 1984, Geohydrology of rocks penetrated by test well UE-25p#1, Yucca Mountain area, Nye County, Nevada: U.S. Geological Survey Water Resources Investigations Report 84-4248, 57 p.

Dailey, W., Lin, W., and Buscheck, T., 1987, Hydrologic properties of Topopah Springs Tuff: Lab measurements: Journal of Geophysical Research, v. 92, no. B8, p. 7854–7864.

Daniels, W.R., Wolfsberg, A., and Rundberg, R.S., 1982, Summary report on the geochemistry of Yucca Mountain and environs: Los Alamos National Laboratory Report LA-9328-MS, 364 p.

Darcy, H., 1856, The public fountains of Dijon, Victor Dalmont, Paris (trans. P. Bobeck, 2004): Dubuque, Iowa, Kendall/Hunt Publishing Co., 506 p.

Driscoll, F.G., 1986, Groundwater and wells (2nd edition): St. Paul, Minnesota, Johnson Division, 1089 p.

Flint, L.E., 1998, Characterization of hydrogeologic units using matrix properties, Yucca Mountain, Nevada: U.S. Geological Survey Water Resources Investigations Report 97-4243, 64 p.

Flint, L.E., and Flint, A.L., 1990, Preliminary permeability and water-retention data for nonwelded and bedded tuff samples, Yucca Mountain area, Nye County, Nevada: U.S. Geological Survey Open-File Report 90-569, 57 p.

Foster, S.S.D., Ellis, A.T., Locilla-Penon, M., and Rodriquez-Estrada, H.V., 1985, Role of volcanic tuffs in ground-water regime of Valle Central, Costa Rica: Ground Water, v. 23, no. 6, p. 795–801.

Fu, L., Milliken, K.L., and Sharp, J.M., Jr., 1994, Porosity and permeability variations in fractured and liesegang-banded Breathitt sandstones (middle Pennsylvanian), eastern Kentucky: Diagenetic controls and implications for modeling dual porosity systems: Journal of Hydrology, v. 154, p. 351–381, doi: 10.1016/0022-1694(94)90225-9.

Fuller, C.M., and Sharp, J.M., Jr., 1992, Permeability and fracture patterns in extrusive volcanic rocks: Implications from the welded Santana Tuff, Trans-Pecos Texas: Geological Society of America Bulletin, v. 104, p. 1485–1496, doi: 10.1130/0016-7606(1992)104<1485:PAFPIE>2.3.CO;2.

Greene, R.T., Rice, G., and Meyer-James, K.A., 1995, Hydraulic characterization of hydrothermally altered Nopal tuff: U.S. Nuclear Regulatory Commission Office of Nuclear Regulatory Research, Division of Regulatory Applications, Center for Nuclear Waste Regulatory Analyses (Southwest Research Institute) NUREG/CR-6356, CNWRA 94–027, variously paginated.

Gregory, J.L., 1981, Volcanic stratigraphy and K-Ar ages of the Manuel Benavides area, northeastern Chihuahua, Mexico, and correlation with the Trans-Pecos volcanic province [Master's thesis]: Austin, University of Texas at Austin, 122 p.

Heiken, G., Goff, F., Gardner, J.N., Baldridge, W.S., Hulen, J.B., Nielson, D.L., and Vaniman, D.T., 1990, The Valles/Toledo caldera complex, Jemez Volcanic Field: New Mexico: Annual Review of Earth and Planetary Sciences, v. 18, p. 27–53.

Henry, C.D., McDowell, F.W., Price, J.G., and Smyth, R.C., 1986, Compilation of K-Ar ages of Tertiary igneous rocks, Trans-Pecos Texas: Austin, University of Texas at Austin, Bureau of Economic Geology, Geological Circular 86–2, 34 p.

Hubbert, M.K., 1956, Darcy's law and the field equations of flow of underground fluids: Transactions of the American Institute of Mining and Metallurgical Engineers, v. 207, p. 222–239.

Kearl, P.M., Zinkl, R.J., Dexter, J.J., and Cronk, T., 1990, Air permeability measurements of the unsaturated Bandelier Tuff near Los Alamos, N.M.: Journal of Hydrology, v. 117, p. 225–240, doi: 10.1016/0022-1694(90)90094-E.

Klinkenberg, L.J., 1941, The permeability of porous media to liquids and gases: Washington, D.C., American Petroleum Institute Drilling production practices, 200 p.

Kruseman, G.P., and de Ridder, N.A., 1990, Analysis and evaluation of pumping test data (2nd edition): Wareningen, The Netherlands, International Institute for Land Reclamation and Improvement, 377 p.

Lahoud, R.G., Lobmeyer, D.H., and Whitfield, M.S., Jr., 1984, Geohydrology of volcanic tuff penetrated by test well UE-25b#1, Yucca Mountain, Nye County, Nevada: U.S. Geological Survey Water Resources Investigations Report 84-4253, 44 p.

LeCain, G.D., 1997, Air-injection testing in vertical boreholes in welded and nonwelded tuff, Yucca Mountain, Nevada: U.S. Geological Survey Water Resources Investigations Report 96-4262, 33 p.

Lipman, P.W., Prostka, H.J., and Christiansen, R.L., 1972, Cenozoic volcanism and plate tectonic evolution of the western United States. I: Early and middle Cenozoic: Royal Society of London Philosophical Transactions, series A, v. 271, p. 217–248.

Lobmeyer, D.H., 1986, Geohydrology of rocks penetrated by test well USW G-4, Yucca Mountain, Nye County, Nevada: U.S. Geological Survey Water Resources Investigations Report 86-4015, 38 p.

Montazer, P., and Wilson, W.W., 1984, Conceptual hydrologic model of flow in the unsaturated zone, Yucca Mountain, Nevada: U.S. Geological Survey Water Resources Investigations Report 84-4345, 55 p.

Moyer, T.C., Geslin, J.K., and Flint, L., 1996, Stratigraphic relations and hydrologic properties of the Paintbrush Tuff nonwelded (PTn) hydrologic unit, Yucca Mountain, Nevada: U.S. Geological Survey Open-File Report 95-397, 151 p.

Neeper, D.A., and Gilkeson, R.H., 1996, The influence of topography, stratigraphy, and barometric venting on the hydrology of unsaturated Bandelier Tuff, Jemez Mountains Region, in Goff, F., et al., eds., New Mexico Geological Society Guidebook no. 47: Socorro, New Mexico, New Mexico Geological Society, p. 427–432.

Peters, R.R., and Klavetter, E.A., 1988, A continuum model for water movement in an unsaturated rock mass: Water Resources Research, v. 24, no. 3, p. 416–430.

Peters, R.R., Klavetter, E.A., Hall, I.J., Blair, S.C., Heller, P.R., and Gee, G.W., 1984, Fracture and matrix hydrologic characterization of tuffaceous materials from Yucca Mountain, Nye County, Nevada: Sandia National Laboratories Report SAND84-1471, 63 p., 5 appendices.

Peterson, D.W., 1979, Significance of the flattening of pumice fragments in ash-flow tuffs, in Chapin, C.E., and Elston, W.E., eds., Ash-Flow Tuffs: Geological Society of America Special Paper 180, p. 195–204.

Peterson, N.P., 1962, Geology and ore-deposits of the Globe-Miami district, Arizona, U.S.: U.S. Geological Survey Professional Paper 342, 151 p.

Pettijohn, F.J., Potter, P.E., and Siever, R., 1987, Sand and sandstone: New York, Springer-Verlag, 553 p.

Rasmussen, T.C., Evans, D.D., Sheets, P.J., and Blanford, J.H., 1993, Permeability of the Apache Leap Tuff: Borehole and core measurements using water and air: Water Resources Research, v. 29, no. 7, p. 1997–2006, doi: 10.1029/93WR00741.

Rautman, C.A., Flint, L.E., Flint, A.L., and Istok, J.D., 1995, Physical and hydrologic properties of outcrop samples from a nonwelded to welded tuff transition, Yucca Mountain, Nevada: U.S. Geological Survey Water Resources Investigations Report 95-4061, 28 p.

Rodgers, D.B., and Gallaher, B.M., 1995, The unsaturated hydraulic characteristics of the Bandelier Tuff: Los Alamos National Laboratory Report LA-12968-MS, 76 p.

Ross, G.S., and Smith, R.L., 1961, Ash-flow tuffs: Their origins, geological relations and identification: U.S. Geological Survey Professional Paper 366, 56 p.

Sara, M.N., 2003, Site assessment and remediation handbook (2nd edition): Boca Raton, Florida, Lewis Publishers, 944 p.

Scott, R.B., Spengler, R.W., Diehl, S., Lappin, A.R., and Chornack, M.P., 1983, Geological character of tuffs in the unsaturated zone at Yucca Mountain, Southern Nevada in Mercer, J.W., Rao, P.S.C., and Marine, I.W., eds., Radioactive and Hazardous Waste Disposal: Ann Arbor, Michigan, Ann Arbor Science, p. 289–335.

Sharp, J.M., Jr., Fu, L., Cortez, P., and Wheeler, E., 1994, An electronic mini-permeameter for use in the field and laboratory: Ground Water, v. 32, no. 1, p. 41–46, doi: 10.1111/j.1745-6584.1994.tb00609.x.

Sharp, J.M., Jr., Robinson, N.I., Smyth-Boulton, R.C., and Milliken, K.L., 1995, Fracture skin effects in groundwater transport, in Rossmanith, H.-P., ed., Mechanics of jointed and faulted rock: Proceedings of the Second International Conference on the Mechanics of Jointed and Faulted Rock, Vienna, Austria, 10–14 April 1995, p. 449–454.

Shenhav, H., 1971, Lower Cretaceous sandstone reservoirs, Israel: Petrography, porosity, and permeability: American Association of Petroleum Geologists (AAPG) Bulletin, v. 55, no. 12, p. 2194–2224.

Smith, R.L., 1960, Zones and zonal variations in welded ash flows: U.S. Geological Survey Professional Paper 354F, 22 p.

Smith, R.L., and Bailey, R.A., 1966, The Bandelier Tuff: A study of ash flow eruption cycles from zoned magma chambers: Bulletin of Volcanology, v. 29, p. 83–104.

Smyth-Boulton, R.C., 1995, Porosity and permeability controls in the Santana ash-flow tuff, Trans-Pecos Texas [M.A. thesis]: Austin, Department of Geological Sciences, The University of Texas at Austin, 100 p.

Thoma, S.G., Gallegos, D.P., and Smith, D.M., 1992, Impact of fracture coatings on fracture/matrix flow interactions in unsaturated, porous media: Water Resources Research, v. 28, no. 5, p. 1357–1367, doi: 10.1029/92WR00167.

Thordarson, W., Rush, F.E., and Waddell, R.K., 1985, Geohydrology of test well USW H-3, Yucca Mountain, Nye County, Nevada: U.S. Geological Survey Water Resources Investigations Report 84-4272, 38 p.

Weeks, E.P., and Wilson, W.E., 1984, Preliminary evaluation of hydrologic properties of cores of unsaturated tuff, test well USW H-1, Yucca Mountain, Nevada: U.S. Geological Survey Water Resources Investigations Report 84-4193, 30 p.

Whitfield, M.S., Jr., Eshom, E.P., Thordarson, W., and Schaefer, D.H., 1985, Geohydrology of test well USW H-4, Yucca Mountain, Nye County, Nevada: U.S. Geological Survey Water Resources Investigations Report 85-4030, 33 p.

Winograd, I.J., 1971, Hydrogeology of ash-flow tuffs: A preliminary statement: Water Resources Research, v. 7, no. 4, p. 994–1006.

Wohletz, K., and Heiken, G., 1992, Volcanology and geothermal energy: Berkeley, California, University of California Press, 432 p.

MANUSCRIPT ACCEPTED BY THE SOCIETY 29 DECEMBER 2005

Geological Society of America
Special Paper 408
2006

Chapter 4.2

Hydrogeology of the city of Rome

A. Corazza

Dipartimento di Protezione Civile, Rome, Italy, Via Ulpiano 11, 00193

G. Giordano
D. de Rita
Universita degli Studi di Roma—TRE, Rome, Italy

HYDROGEOLOGIC FRAMEWORK

The hydrogeologic framework of the area around Rome is characterized by a Pliocene–Lower Pleistocene marine claystone bedrock, which is the major regional aquiclude underlying the shallow hydrogeologic units; the claystone bedrock is hundreds of meters thick and has a very low permeability (Ventriglia, 1971, 1990, 2002; Albani et al., 1972; Boni et al., 1988; Corazza and Lombardi, 1995a; Funiciello and Giordano, 2005; Capelli et al., 2005). The nearly impermeable bedrock is overlain by Lower to Middle Pleistocene marine to continental sediments (claystones, sandstones, and thick sequences of conglomerates), which are in turn overlain by and partly interfingered with Middle to Upper Pleistocene volcanic deposits from the Sabatini volcanic complex to the north and the Colli Albani volcanic complex to the south (Funiciello and Giordano, 2005; Capelli et al., 2005). Holocene alluvial sediments cap the stratigraphic sequence along the present-day river systems (Corazza et al., 1999; Funiciello and Giordano, 2005). Holocene eolian sand dunes cover a narrow area along the present-day Tyrrhenian coastline.

All rock sequences overlying the claystone bedrock are aquifers, the geometry and circulation of which are controlled by both the evolution with time of the paleotopographic setting and the vertical and lateral variations of lithologies, each with different permeabilities (Capelli et al., 2005).

Other locally important hydrogeologic units, only present in the urbanized areas, are the backfill deposits accumulated during 3000 yr of human civilization in the Roman area. On a regional scale, the Pliocene–Lower Pleistocene marine claystone aquiclude overlies a deep aquifer in highly deformed Mesozoic-Cenozoic carbonates (Boni et al., 1988), which is recharged from the

Apennine region. Tectonic and volcano-tectonic discontinuities control groundwater flow on a local scale, as well as gas and fluid leakage from the deeper Mesozoic-Cenozoic carbonate reservoir, as evidenced by the presence of several low- to medium-enthalpy hydrothermal springs (e.g., Tivoli) (Funiciello et al., 2003; Carapezza et al., 2003; Tuccimei et al., 2006).

HYDROGEOLOGY OF THE VOLCANIC DEPOSITS

Major aquifers around Rome can be found in interbedded pyroclastic deposits (ignimbrites, surge and fall deposits, and reworked pyroclastic materials) and lavas. The volcanic aquifers are characterized by good permeability as a whole (primary permeability in tuffs and secondary in lavas), although considerable variation in permeability may be found among different deposits (Capelli et al., 2000, 2005). Fractured lavas embedded in generally lower-permeability tuffs represent preferential groundwater drainage pathways.

The volcanic deposits host many aquifers at different depths. The main aquifer is located at the base of the volcanic pile, and is confined by either the oldest low-permeability volcanic products or by the Pliocene-Pleistocene marine sediments. Above the main aquifer, many local perched aquifers can be found confined at the base by low-permeability pyroclastics and thick paleosols. Aquifers are generally unconfined because of the lateral discontinuity of impermeable layers, although confined aquifers, especially thick lava-flow units, can be found locally (Capelli, et al., 2005).

To the east and south of the Tiber River, volcanic deposits belong to the Colli Albani volcanic complex, and the main aquifer flows radially outward from the volcano reaching the southern

Corazza, A., Giordano, G., and de Rita, D., 2006, Hydrogeology of the city of Rome, *in* Heiken, G., ed., Tuffs—Their properties, uses, hydrology, and resources: Geological Society of America Special Paper 408, p. 113–118, doi: 10.1130/2006.2408(4.2). For permission to copy, contact editing@geosociety.org. ©2006 Geological Society of America. All rights reserved.

and eastern suburbs of Rome. The river system and springs in the area, which supply much of the potable water in the Roman area, gain water from the main aquifer. Examples include the Acqua Vergine spring, which presently has a discharge of 600 L/s (and was 1200 L/s during classical Roman times) and drains water from lavas, or the many springs utilized by the Appio and Augusto aqueducts of the ancient Romans, which discharged more than 1000 L/s (presently ~600 L/s). In the city center, a number of small springs are fed by water from a localized shallower aquifer (important for the local history and economy) (Coppa et al., 1984; Pisani Sartorio and Liberati, 1986; Corazza and Lombardi, 1995b). North of the Aniene River, some springs with few liters per second of discharge gain water from the aquifer that flows toward the Tiber and Aniene Rivers.

West of the Tiber River, volcanic deposits mainly belong to the Sabatini volcanic complex, and thin southward toward Rome. The aquifer flows radially away from the volcanic complex, i.e., from the northwest (where it reaches the surface at the Bracciano and Martignano lakes) to the south-southeast (Boni et al., 1988; Ventriglia, 1990; Capelli et al., 2005). Some important springs are located in the northern sector of the Roman area, where the volcanic pile is thickest and the main aquifer is more substantial. Smaller springs (<0.5 L/s) are related to perched, local aquifers. Many water wells have been drilled in the area that drain water from the main aquifer, with resulting specific capacity of up to tens of liters per meter of water-table drawdown.

ANCIENT SPRINGS OF ROME

The presence of water under Rome and its use from springs and water wells may have been one of the main factors influencing the siting of the early city. During classical Roman time, the many springs within the city and the surrounding area were well known and utilized (Lanciani, 1881; Pisani Sartorio and Liberati, 1986; Corazza and Lombardi, 1995a, 1995b). Some of them still exist, and they are an invaluable archaeologic and natural treasure to be preserved. These springs, together with water wells and the Tiber River, were the only available water in Rome for a long time. Water catchment structures are found in the city center, as well as in the surrounding areas, confirming the abundance of water in this territory during ancient times.

Collection of water from local springs and water wells was abandoned after the Romans began to build aqueducts, and some of the springs were transformed into sacred sites. Most of the Roman engineering patrimony was lost with the fall of the Roman Empire. The barbarian Vitige knocked down the aqueducts in 537 A.D., and the city was forced to rely on local springs and old catchments, although most were unusable or backfilled with manmade debris. Between the sixteenth and the seventeenth centuries, searches for new sources of groundwater were sponsored by the papal administrations. Springs (active presently or in the past) that gain(ed) water from volcanic deposits, mostly tuffs, are listed below:

Quirinale Hill

Acque Sallustiane. The Sallustiane springs were located in the valley between the Pincio and Quirinale hills, and flowed into the Amnis Petroniae stream. The springs gained water both from the volcanic deposits and the underlying Pleistocene conglomerates, which cropped out at the base of the valley, but are presently completely buried by debris.

Acqua di S. Felice. The S. Felice fountain, which is located in the gardens of the Presidential Quirinale Palace, receives its water from a small spring that is likely the ancient Fons Cati.

Acque Fontinali-Fonte del Grillo. Some springs collectively known as Acque Fontinalis outcropped along the southern slope of Quirinale hill, where the Fontinalis door of the Serviane walls was located.

Palatino Hill

Lupercale. Water from this spring was renowned during Roman time, as it was consecrated to Lupercus Faunus (the one who keep wolves away) and was used during ceremonies of the Lupercalia feasts. It is known to be the first spring used for sacred ceremonies in ancient Rome. The location of the spring is uncertain, but the geology of Colle Palatino suggests that it emerged from the Tufo Lionato formation, a zeolitized ignimbrite from the Colli Albani volcano.

Fonte di Pico. The Fonte di Pico was named by Ovidius and was located in a cave at the base of Colle Palatino, facing the Tiber River. It probably disappeared when the present-day river embankments were built up during late nineteenth century. This spring probably derived its water from the volcanic deposits, which were, in turn, drained by Pleistocene sandy-travertine deposits.

Caffarella Valley

Fonte Egeria. The Caffarella Valley was renowned for the many mineralized springs and their therapeutic properties. The mineralization was the product of mixing of groundwater with upwelling, deep-seated fluids that rose in this area along volcano-tectonic fractures. The Fonte Egeria is known to have been used by Erode Attico for a Nymphaeum in his villa. The ruins of the Nymphaeum still host a spring of mineralized water.

Area of Salone

Acqua Vergine. The Acqua Vergine is composed of a number of springs located ~10 km to the east of Rome, along the old Via Collatina, in a once swampy area. The legend by Frontinus and Plinio il Vecchio tells of a young virgin girl who indicated the spring's location to some soldiers. More likely, the origin of the name is to be ascribed to the purity of the water, which Marziale defined as gelida (icy), nivea (snowy), and cruda (sharp) to underline its freshness. The Acqua Vergine was captured by the Romans and transported into Rome via the Virgo Aqueduct, the only aqueduct of Roman time that is still functioning. The aqueduct was designed by Agrippa and inaugurated on the 9th of June, 19 B.C. Frontinus states that the aqueduct discharge was 101,000 m³/d, versus the present-day discharge of 78,000 m³/d.

PERMEABILITY OF VOLCANIC DEPOSITS

The permeabilities of volcanic deposits vary as a function of primary porosity, compaction processes, sealing, and secondary fracturing. Lava flows are generally characterized by very low matrix porosities, but are intensely fractured, either by columnar jointing or by postdepositional faulting and fracturing. Lavas have well-developed fracture systems, which result in an effective drainage system via fracture permeability. Lava flows near Rome are generally confined to paleovalleys, where they are locally up to few tens of meters thick, but lack lateral continuity. To the south-southeast of the city, a lava plateau, up to 100 m thick, has been revealed from drillings; it constitutes a major drainage pathway in that region (Funiciello and Giordano, 2005).

Pyroclastic deposits, on the other hand, show large variations in permeability. The porosity of such deposits varies as a function of granulometry, which depends on the type and intensity of explosive processes and on the mechanism of their emplacement, as well as postdepositional alteration processes.

Fallout deposits are generally well sorted and highly permeable, at least in proximal and medial areas where block- to lapilli-sized deposits are common. Fallout deposits also mantle pre-existing topography and therefore are laterally continuous, if not eroded.

Pyroclastic flow deposits emplaced by density currents (i.e., ignimbrites and surge deposits) are generally poorly sorted and have a wider range of particle sizes, with abundant fine ash (<0.064 mm). Ignimbrites and surge deposits have generally low to medium permeabilities. Individual surge deposits are also relatively thin (up to few meters) and are confined to proximal areas of the volcanic center, whereas individual ignimbrites can be up to few tens of meters thick and are widely dispersed around the volcanic centers of the Colli Albani and Sabatini volcanic complexes. The ignimbrites are massive, with pumice, scoria, and lithic clasts dispersed throughout an ashy matrix. To the south and east of the city of Rome, the main aquifer is mainly represented by the "Pozzolane rosse"–"Pozzolane nere" ignimbrite sequence (Capelli et al., 2000). The permeability of these ignimbrites, however, varies as a function of postemplacement lithification processes. For example, zeolitization is characteristic of ignimbrite units in the Roman area, including the "Tufo Lionato" from the Colli Albani (Watkins et al., 2002) and the "Tufo Rosso a Scorie Nere" from the Sabatini volcanic field. Zeolitized ignimbrites can be local aquifers if they develop a good fracture network.

The alteration of volcanic glass to halloysite may reduce the primary porosity of pyroclastic deposits, filling up fractures in both lavas and zeolitized ignimbrites. Permeabilities may also be reduced by deposition of carbonates from circulating water, especially by travertine where hydrothermal fluids mix with groundwater (Tuccimei et al., 2006).

Permeability values (cm/s) have been determined for volcanic deposits around Rome with the Lefranc test at variable head (Table 4.2.1). Transmissivity values of aquifers in volcanic terrains (m²/s × 10⁻³) were obtained by pumping tests (Table 4.2.2).

MINERAL WATERS

The Roman area is characterized by a wide range of mineral waters, generally cold, but occasionally >20 °C (ipothermal waters) (Camponeschi and Nolasco, 1982; Brondi et al., 1995). There are six plants for mineral water extraction and bottling in Rome (Tables 4.2.3 and 4.2.4), and this is probably a unique case in the world for a metropolis. The mineralization is caused in part by normal cationic enrichment and partly by mixing of shallow groundwater with deep-seated hydrothermal fluids upwelling along tectonic fractures.

Chemical and Physical Properties of Groundwater

Groundwater in the Roman area is classified as bicarbonated-alkaline-earth, with the noticeable exception of mineralized waters (Table 4.2.5).

Vulnerability of the Aquifers to Pollution

The Roman area, like any urbanized and industrialized area, is characterized by the pollution of groundwater. Potential pollutants include liquid and gaseous industrial waste, hydrocarbon depots (comprising petrol stations), solid waste depots (either controlled or uncontrolled), car parks, cesspools, injection wells, areas lacking sewage systems, and farming areas where there is potential pollution by fertilizers, pesticides, herbicides, and antiparasites. Extensive urbanization has also considerably altered soil and surface conditions, increasing aquifer vulnerability. Construction has also introduced pollutants into the aquifers, as well as reduced the recharge area (the urbanized areas are ~20% of the Rome Council area), therefore depleting the aquifer and its self-purification properties.

As no specific work on the vulnerability of volcanic aquifers in the Roman area has been carried out yet, we refer to the general knowledge of an aquifer to assess its vulnerability to pollution (Table 4.2.6). Unconfined aquifers in the Roman area are generally vulnerable because they are shallow and lack a significant thickness of impermeable rocks; pollutants can easily reach the aquifers. Confined aquifers are embedded in, and therefore protected by, impermeable rocks. However, the large number of water wells and catchment basins present in the area has induced an artificial hydraulic continuity between aquifers at different levels, so that confined aquifers are virtually absent, and pollutants leak from upper aquifers to deeper levels. Data from the Roman Council (Table 4.2.7) show that none of the aquifers

TABLE 4.2.1. PERMEABILITY OF VOLCANIC TERRAINS—LEFRANC TESTS

Lithology	Number of tests	Permeability (cm/sec)				
		mean	median	mode	minimum	maximum
pozzolan	24	0.002213	0.0002	0.00004	0.00001	0.03
altered tuffs	7	0.002039	0.000405	-	0.000029	0.00663
zeolitized tuffs	6	0.006081	0.002249	-	0.000052	0.0232

TABLE 4.2.2. TRANSMISSIVITY OF VOLCANIC AQUIFERS—PUMPING TESTS

Aquifer	Number of tests	Transmissivity (m^2/sec x 10^{-3})				
		mean	median	mode	minimum	maximum
pozzolan	18	2.0	2.0	2.0	0.1	6.0
lava	27	25.0	5.5	0.2	0.2	200.0
zeolitized tuffs	6	0.3	2.5	2.0	0.6	8.0
tuffs	16	1.3	0.6	0.6	0.08	5.4

TABLE 4.2.3. MINERAL WATERS OF ROME

Name	Water capture	Hydrogeologic complex	Classification
Acqua Sacra	well	Pleistocene deposits	M - F
Acqua Santa Sorgente Ninfa Egeria	springs	volcanic products	MM - F
Acqua Appia	wells	volcanic products	MM - F
Acqua Santa Maria delle Capannelle	wells	volcanic products	MM - F
Acqua Acetosa San Paolo	wells and springs	Pleistocene deposits	M - I
Acqua Laurentina	wells	Pleistocene deposits	M - I

Note: MM—Medium mineral (dry residue: 200–1000 mg/l); M—Mineral (dry residue: >1000 mg/l); F—Cold (T < 20 °C); I—Ipothermal (*T* > 20 °C)

TABLE 4.2.4. CHEMICAL AND PHYSICAL CHARACTERISTICS OF MINERAL WATERS OF ROME

Name	Dry residue (g/l)	T (°C)	pH	Hardness (°F)	Na (mg/l)	K (mg/l)	Ca (mg/l)	Mg (mg/l)	Cl (mg/l)	SO_4 (mg/l)	HCO_3 (mg/l)
Acqua Sacra	1.03	16.5	6.40	70	68.0	59.0	232.5	29.2	39.1	68.1	973.3
Acqua Santa Ninfa Egeria	0.64	17.9	-	34	45.0	53.0	94.5	24.1	31.5	30.0	506.3
Acqua Appia	0.70	18.0	6.02	50	43.0	48.2	125.0	20.1	34.8	30.5	549.0
Acqua S.M. delle Capannelle	0.69	18.0	6.10	45	42.8	39.0	116.1	38.0	31.9	27.9	609.8
Acqua Acetosa San Paolo	2.27	20.0	6.00	113	254.0	242.0	333.0	70.1	50.2	273.1	1970.3
Acqua Laurentina	1.31	20.5	6.30	78	80.0	120.4	245.8	40.1	35.5	115.3	1238.3

TABLE 4.2.5. CHEMICAL AND PHYSICAL CHARACTERISTICS OF GROUNDWATERS IN THE ROMAN AND COLLI ALBANI AREAS

Parameter	Samples														
	1	2	3	4	5	6	7	8	9	10	11	12	13	14	15
T (°C)	12.6	21.2	11.2	13.3	12.0	10.4	5.1	14.2	17.1	13.3	14.0	11.2	16.9	17.5	17.9
pH	6.9	6.4	6.5	66	5.7	3.9	5.85	7.2	5.3	6.5	8.2	2.8	8.8	6.5	7.3
Eh (mV)	355	384	182	398	518	405	444	392	148	381	430	486	391	541	554
El. cond. (μS cm^{-1})	690	1916	2766	693	825	1500	890	349	712	846	1480	8455	1046	778	649
Ca (meq L^{-1})	3.50	12.00	29.00	3.10	3.50	9.30	2.60	1.20	1.50	4.00	0.90	20.00	0.60	2.50	1.50
Mg (meq L^{-1})	0.95	3.60	6.30	1.00	1.90	4.10	1.40	0.79	2.30	1.90	0.17	10.00	0.24	1.05	0.53
Na (meq L^{-1})	0.73	4.00	0.36	1.10	1.10	1.90	1.20	0.49	1.30	1.30	8.50	1.60	5.01	1.63	2.10
K (meq L^{-1})	0.57	2.90	0.12	0.88	1.80	6.80	0.95	0.35	1.00	1.10	3.40	3.70	1.98	1.47	1.32
HCO$_3$ (meq L^{-1})	5.40	20.00	12.00	5.20	7.20	0.00	5.00	1.90	2.20	7.50	3.20	0.00	3.19	6.20	4.91
SO$_4$ (meq L^{-1})	0.35	1.80	24.00	0.64	0.05	18.00	0.67	0.11	3.40	0.31	7.60	150.0	3.90	0.74	0.37
Cl (meq L^{-1})	0.35	0.86	0.33	0.49	0.32	0.58	0.37	0.58	1.70	0.48	1.70	0.59	1.08	0.35	0.70
SiO$_2$ (ppm)	43.1	53.3	20.06	44.5	73.5	74.0	52.5	43.2	109.9	53.9	9.9	114.0			
F (ppm)						4.00							5.00	2.10	3.30
B (ppm)	0.09	1.20	0.16	0.13	0.15	0.18	0.10		0.11		1.90	1.10			
Sr (ppm)	0.54	1.93	8.79	0.63	1.14	1.90	0.74	0.43	0.69	1.20	0.28	2.53	0.13	0.82	0.61
Li (ppb)	3.9	167.2	31.5	8.1	2.7	14.0	2.3	0.6	25.0	3.2	13.0	320.0	21.9	29.9	15.9
Rb (ppb)	28.0	130.0	13.0	25.0	56.0	270.0	49.0	83.0	560.0	77.0	158.0	2400			
Cs (ppb)	2.0	3.0	2.0	2.0	2.0	12.0	3.0	1.0	4.0	1.0	2.0	190.0			
As (ppb)		27.0	0.0		2.2	0.4			2.2		12.2	4.3	0.1	0.1	0.1
Rn$_{(a)}$ (pCi L^{-1})	648	295	378	900	1148		1404	1950	2340	2870	1656	1224	1077	1239	995
Rn$_{(g)}$ (pCi L^{-1})						3100									
He$_{(a)}$ (ppb v/v)		166	667	109	218		96	72	966	230	238	251	70	194	79
He$_{(g)}$ (ppb v/v)						3000						5000			
PCO$_2$ (mbar)	37	414	179	71	769		380	<10	556	127	<10		<10	102	<10

Samples: 1—Acqua Felice well; 2—Acqua Laurentina well; 3—Acqua Solf a spring; 4—Acqua Vergine spring; 5—Barozze 1 well; 6—Cava dei Selci well (gas); 7—Facciata di Nemi spring; 8—Fonte Vetrice spring; 9—Grotta Dauni well; 10—Squarciarelli spring; 11—Valle Ariccia well; 12—Zolforata well (gas); 13—Sabatini well; 14—Ariccia well 113; 15—Asveca well.

TABLE 4.2.6. VULNERABILITY
OF VOLCANIC AQUIFERS NEAR ROME

Aquifer	Vulnerability degree
Fractured lithoid zeolitized ignimbrites and lavas	high
Multilayered aquifers in nonindurated tuffs	medium–low

TABLE 4.2.7. MICROBIOLOGICAL CHARACTERISTICS OF GROUNDWATERS OF ROME

Water capture (depth)	Samples	C.F. = 0 (%)	C.F. = 1–10 (%)	C.F. = 11–50 (%)	C.F. = >50 (%)
Springs	50	40	26	18	16
Wells (1–10 m)	49	47	29	4	20
Wells (11–20 m)	60	47	13	20	20
Wells (21–30 m)	71	59	18	8	14
Wells (31–40 m)	81	60	27	7	5
Wells (41–55 m)	77	65	25	6	4
Wells (56–75 m)	71	76	10	8	6
Wells (75–100 m)	31	58	16	16	10
Wells (>100 m)	12	83	17	0	0
Wells (unknown depths)	36	58	33	8	0
Total	541	59	21	10	10

in the area are protected from microbiological pollution, not even the deepest ones.

REFERENCES CITED

Albani, R., Lombardi L., and Vicinanza, P., 1972, Idrogeologia della Città di Roma: Roma, Ingegneria Sanitaria, v. 20, no. 3, 38 p.

Boni, C., Bono, P., and Capelli, G., 1988, Carta idrogeologica della regione Lazio: Roma, Regione Lazio, Università degli Studi di Roma "La Sapienza," scale 1:250,000.

Brondi, M., Campanile, R., Dall'Aglio, M., Orlandi, C., Tersigni, S., and Venanzi, G., 1995, Acque naturali, *in* ENEA (Ente Nazionale Energia e Ambiente): Lazio Meridionale: Sintesi delle Ricerche Geologiche Multidisciplinari: ENTA, Dipartimento Ambiente, Studi e Ricerche Series, 350 p.

Camponeschi, B., and Nolasco, F., 1982, Le risorse naturali della regione Lazio: Roma e i Colli Albanizz: Roma, Regione Lazio, Tipolitografia Edigraf, v. 7, 547 p.

Capelli, G., Mazza, R., Giordano, G., Cecili, A., De Rita, D., and Salvati, R., 2000, The Colli Albani volcano (Rome, Italy): Breakdown of the equilibrium of a hydrogeological unit as a result of unplanned and uncontrolled over-exploitation: Hydrogéologie, v. 4, p. 63–70.

Capelli, G., Mazza, R., and Gazzetti, C., eds., 2005, Strumenti e strategie per la tutela e l'uso compatibile della resorsa idrica nel Lazio—Gli acquiferi vulcanici: Bologna, Pitagora Editrice Bologna, 186 p. and maps.

Carapezza, M.L., Badalamenti, B., Cavarra, L., and Scalzo, A., 2003, Gas hazard assessment in a densely inhabited area of Colli Albani volcano (Cava dei Selci, Roma): Journal of Volcanology and Geothermal Research, v. 123, p. 81–94, doi: 10.1016/S0377-0273(03)00029-5.

Coppa, G., Pediconi, L., and Bardi, G., 1984, Acque e acquedotti a Roma 1870–1894: Roma, Quasar, 233 p.

Corazza, A., and Lombardi, L., 1995a, Idrogeologia del centro storico, *in* Funiciello, R., ed., La geologia di Roma: Memorie Descrittive Carta Geologica d'Italia, v. 50, p. 173–211.

Corazza, A., and Lombardi, L., 1995b, Le acque sotterranee, *in* Cignini, B., Massari, G., and Pignatti, S., eds., L'ecosistema Roma:Amiente e territorio: Conoscenze attuali e prospettive per il duemila: Roma, Fratelli Palombi, p. 40–46.

Corazza, A., Lanzini, M., Rosa, C., and Salucci, R., 1999, Caratteri stratigrafici, idrogeologici e geotecnici delle alluvioni tiberine nel settore del Centro Storico di Roma: Italian Journal of Quaternary Sciences (Il Quaternario), v. 12, no. 2, p. 215–235.

Funiciello, R., and Giordano, G., 2005, Carta geologica del comune di Roma, Vol. 1: Dipartimento Scienze Geologiche, Università di Roma TRE–Comune di Roma–APAT, 18 maps, 3 cross-sections, scale 1:10,000. Open file: http://host.uniroma3.it/laboratori/labgis/cartaroma/.

Funiciello, R., Giordano, G., and De Rita, D., 2003, The Albano maar lake (Colli Albani volcano, Italy): Recent volcanic activity and evidence of pre-Roman age catastrophic lahar events: Journal of Volcanology and Geothermal Research, v. 123, no. 1–2, p. 43–61, doi: 10.1016/S0377-0273(03)00027-1.

Lanciani, R., 1881, Topografia di Roma antica. I commentarii di Frontinus intorno le acque e gli acquedotti: Memoirie Regia Accademia Lincei, ser. 3, v. 4, p. 215–614.

Pisani Sartorio, G., and Liberati, A.S., 1986, Il Trionfo dell'Acqua, Exhibition Catalogue: Roma, Peleani publishing house, 195 p.

Tuccimei, P., Giordano, G., and Tedeschi, M., 2006, CO_2 release variations during the last 2000 years at the Colli Albani volcano (Roma, Italy) from speleothems studies: Earth and Planetary Science Letters, v. 243(3-4), p. 449–462.

Ventriglia, U., 1971, La geologia della città di Roma: Roma, Provincia di Roma, 513 p.

Ventriglia, U., 1990, Idrogeologia della Provincia di Roma: Roma, Provincia di Roma, 255 p.

Ventriglia, U., 2002, Geologia del territorio del Comune di Roma: Roma, Amministrazione Provinciale di Roma, 810 p.

Watkins, S.D., Giordano, G., Cas, R.A.F., and De Rita, D., 2002, Emplacement processes of the mafic Villa Senni eruption unit (VSEU) ignimbrite succession, Colli Albani volcano, Italy: Journal of Volcanology and Geothermal Research, v. 118, no. 1–2, p. 173–203, doi: 10.1016/S0377-0273(02)00256-1.

Manuscript Accepted by the Society 29 December 2005

Geological Society of America
Special Paper 408
2006

Chapter 5

Construction in regions with tuff deposits

R. Funiciello
Dipartimento di Scienze Geologiche, Università degli Studi di Roma Tre, Roma, Italy

G. Heiken
331 Windantide Place, Freeland, Washington 98249-9683, USA

R. Levich
Gomoa Ojobi, Ghana

J. Obenholzner
Museum of Natural History, Vienna

V. Petrov
Institute of Geology of Ore Deposits, Russian Academy of Sciences, Moscow, Russia

INTRODUCTION

Wherever there is a natural occurrence of tuff, man has utilized cut slabs or blocks of this stone type for construction or carved caves for shelter. Easy to cut, relatively light, and useable as an insulation, tuffs have a wide range of uses, ranging from tunnels for storage of high-level nuclear waste to Spanish colonial architecture in Guadalajara, Mexico. Even the word "tuff" is derived from "tufo," an Italian quarryman's term for any rock that can be cut with a knife.

Nonwelded or poorly welded massive ignimbrite deposits are commonly used throughout the world for construction. The rock is easily cut and dressed into blocks of almost any shape and is resistant to weathering. The blocks are lighter than most stone of equivalent size, have some insulation value, and are attractive.

In Latin America, tuffs are used frequently for construction, especially for large public buildings. Large buildings in the Peruvian city of Arequipa, known as La Ciudad Blanca ("the White City"), are constructed of nonwelded white ignimbrite altered by vapor-phase activity (Fenner, 1948). Blocks are cut with saws and dressed with hammer and ax; the cooling joints in these ignimbrites are used as block faces when possible. The ignimbrite blocks are also resistant to weathering, with little degradation of tuff blocks occurring in 300-yr-old colonial buildings (Jenks and Goldich, 1956).

In Naples, Italy, the large-volume, partly hydrovolcanic, fine-grained tuffs (mostly massive ignimbrites deposited during caldera-forming events in the Phlegrean Fields) have been a constant source of building stone for thousands of years. The Neapolitan Yellow Tuff (12,000 yr old), Campanian Ignimbrite (37,000 yr old), and younger tuff rings of the Phlegrean Fields cover a large area within and around the city of Naples (Orsi et al., 1996). The first use of tuff in Naples was by the Greeks in 800 B.C., and ever since, the rock has been used to construct churches, castles, fortresses, and apartment and office buildings. A tuff quarry in the Phlegrean Fields was used by the Romans for a reservoir, and tunnels have been used for streets, aqueducts, bomb shelters during World War II, and rubbish dumps. These tuff deposits continue to be quarried today as the basic building stone of the region.

In Rome, Italy, tuff deposits (mostly fine-grained ignimbrites) from the Alban Hills and Sabatini volcanic fields overlap in the center of the city. The famous "Seven Hills" of Rome are actually the eroded ends of a tuff plateau from the Alban Hills volcanic field. From Etruscan times to the present, various facies of these tuff deposits have been used in construction, and the

Funiciello, R., Heiken, G., Levich, R., Obenholzner, J., and Petrov, V., 2006, Construction in regions with tuff deposits, *in* Heiken, G., ed., Tuffs—Their properties, uses, hydrology, and resources: Geological Society of America Special Paper 408, p. 119–126, doi: 10.1130/2006.2408(05). For permission to copy, contact editing@geosociety.org. ©2006 Geological Society of America. All rights reserved.

deposits themselves have offered ideal foundation materials for the city's growth.

FIELD PROPERTIES OF TUFFS TO BE SELECTED AS BUILDING STONES OR FOR UNDERGROUND STRUCTURES

Facies

When selecting a quarry site or a site for an underground structure, the ideal rock is a massive pyroclastic flow deposit that is poorly to moderately welded (Nelson and Anderson, 1992, scale of 3.5–4.0). With the exception of cooling joints and concentrations of lithic clasts, the stone is uniform and has considerable strength. Surge facies should be avoided because of their heterogeneous texture, which results in blocks that are friable and easily weathered. Unconsolidated pumice and ash fallout generally make poor building materials, but are commonly mined (quarried) for use as abrasives or insulating material.

Underground structures are relatively easy to excavate in nonwelded or poorly welded ignimbrite. The famous underground cities of Turkish Cappadocia are a classic example of this use. In the Canary Islands, Spain, and on the Island of Thera, Greece, farmers excavate rooms in ignimbrites with a pick and shovel, resulting in structurally sound homes and stables that are clean, dry, and have even interior temperatures.

Exceptions to the generalizations described for welded or partly welded ignimbrites are those ignimbrites formed in part during hydrovolcanic (involving magma-water interactions) eruptions. These deposits are commonplace throughout the volcanic fields along the Tyrrhenian Sea, Italy, where rising magmas have mixed explosively with water in karstic aquifers that underlie the coastal plains. From the hill towns of Tuscany and Umbria to the Campanian Plain, fine-grained ignimbrites, cemented by clays and zeolites, have been quarried for construction for thousands of years. In these areas, the entire deposit (except surge facies) is ideal for quarrying and construction. The same deposits have also been ideal for development of underground structures—for example, Romans quarried tuffs in Rome and Naples underground because the land was too valuable for buildings and agriculture to be lost to open quarries. Underground stone quarries form immense networks under these cities today and can collapse under modern streets or buildings.

Fractures

Within welded tuffs, there can be problems with quarrying large blocks or excavating underground structures because of the abundance of cooling joints produced during cooling and contraction of the ignimbrite deposit (for a discussion of this phenomenon, see Chapter 2.3 by Wohletz, this volume). Unless the actual joint surfaces are used as block surfaces, these massive tuff facies should be avoided. In certain circumstances, smaller blocks are cut from welded tuffs for use in construction.

This is the case in many towns and cities of the Ethiopian Plateau, but the quarrying is very labor-intensive, with the blocks cut and dressed by hand. Wohletz (this volume) studied jointing patterns in the Bandelier Tuff of New Mexico and found that the most intense joint development is associated with the more strongly welded zones, where the average fracture spacing is 65 fractures per hundred meters of outcrop. In plan view (deposit surface), these fractures form polygonal or rhombohedral patterns. Most are steeply dipping, nearly vertical, but there are some horizontal fractures. The average fracture aperture (opening) is 0.7–1.0 cm, although those near faults can be as much as 5 cm.

The fracture study by Wohletz (this volume) was conducted in a semiarid climate where outcrops are superb. When evaluating a site for quarrying or construction in tuffs (especially ignimbrites), it may be more difficult to study the degree of fracturing. Toshioka et al. (1995) experimented with ground-penetrating radar to evaluate fracturing in tuffs for areas with less than ideal rock exposures.

The purpose of the Toshioka et al. (1995) study was to use ground-penetrating radar to evaluate the distribution and continuity of cracks in rock units at cliff or slope faces to predict rock-fall danger. The authors carried out measurements in a 10-m-high vertical quarry wall in poorly welded tuff (?). Reflected waves within the tuff were detected to a depth of ~4 m. By comparing ground-penetrating radar records with crack condition visible at the rock surface, the waves were most strongly reflected by cracks containing water. The wall was mapped for crack distribution and orientation, and presence of water and/or crack fillings. Profiles were made by using antennas designed to operate at four different frequencies. They used the SIR system 10, manufactured by Geophysical Survey Systems, Inc. Center frequencies were 100, 300, 500, and 900 MHz. Measurements were made using the profiling method, in which the transmitting and receiving antennas were moved continuously, while keeping a set distance. Operating depth is greater with lower-frequency antennas. Reflected waves are generated from cracks containing clay, spring flow, and dripping water. Waves are not generated from closed or dry cracks.

The sophistication and utility of ground-penetrating radar is increasing every year, and this technique could be used routinely for a quick evaluation of tuff sites picked for construction or quarrying in temperate or humid regions. It appears that the technique may not work in arid regions, where cracks are dry.

Fault Zones in Tuff Deposits

In welded or partly welded tuffs, the rock is remarkably brittle and is shattered when broken by faults. In his study of the Bandelier Tuff, Wohletz (1996) found that near faults, fracture spacing increases to 230 per hundred meters of outcrop. Wohletz (1996) also found that fractured tuff accommodates strain incrementally with each fracture and can actually conceal faults and the degree of offset.

Cementation and Fracture Filling

The brief discussion of rock quality, facies, and fractures covered some of the basics about evaluating tuffs for quarrying and foundation sites. The processes of vapor-phase alteration within the deposit during cooling and long-term interaction of the pyroclasts with groundwater can change an unacceptable site into an acceptable site. These postemplacement processes can bond pyroclasts by secondary crystal growth and filling of open pore space with mineral cements, such as zeolites and smectitic clays. In older tuffs, even the largest fractures can be filled and bonded with secondary minerals.

When evaluating a site in which tuffs have been consolidated and bonded by secondary minerals, a brief study should be made of the uniformity of this alteration and the minerals present. For example, the presence of abundant swelling clays would make the site undesirable for development.

Paleosurfaces Underlying a Tuff Deposit

If a tuff deposit is being considered as a quarry, a brief study by a trained geologist is needed in order to consider the geometry of the deposit. Within large-volume ignimbrites associated with caldera-forming eruptions, much of the tuff deposit is thick and widespread, and, if rock qualities are desirable, can usually be considered for large quarries. However, in regions where the pre-eruption topography included deep ravines and valleys and where the volume of tephra erupted was small, there are remarkable lateral variations in thickness and facies of the tuff. In extreme examples, a tuff deposit can be hundreds of meters thick where it has filled a paleovalley, but only a few tens of meters thick (or even absent) on adjacent paleoridges. If the deposit has spread out over what was more or less a flat valley, tuff deposits are more uniform in thickness.

GEOTECHNICAL PROPERTIES OF TUFFS

In the past, much of the geotechnical data on tuffs has been collected and published by civil engineers who had little information on the rock types, degree of alteration, and extent of the deposits. The most systematic geotechnical study of tuffs was in Armenia, by Atzagortsian and Martirosian (1962), but this monograph is very difficult to obtain. Italian investigators published a few geotechnical studies in the geological or engineering literature, but most are limited to specific problems or sites. Although not quantitative, quarrymen have, through experience, a good feel for the relationship between geology and rock quality.

However, over the past 20 yr, the U.S. Department of Energy's Yucca Mountain Project has changed this picture. The proposed high-level nuclear waste repository is located in a complex sequence of tuffs (mostly ignimbrites) in an isolated Tertiary-age volcanic field in southern Nevada. To meet the standards of the U.S. Nuclear Regulatory Commission for site suitability, hundreds of core holes, an exploratory tunnel, and thousands of geotechnical measurements of tuffs have been made. This vast data set has been condensed and is included in this volume as the section on geotechnical properties of tuffs (Chapter 3).

QUARRYING TUFF DEPOSITS

Quarrying Technology

Throughout history, where they are available, tuffs have been excavated or quarried because of the ease with which the stone is cut. In more primitive societies, tuff was cut with stone axes or scrapers, and as metals came in to use, blocks were sawn. In pre-Hispanic Mexico, tuffs and lavas were cut with stone and copper tools, mostly by chipping (Holmes, 1985); pick marks are evident at all stages of the quarrying process, and the ancient quarries contain abundant hammer stones. Quarries were located along bluffs for easy access to outcrops and extraction. The Spanish word "sillar" in Latin America refers to nonwelded or poorly welded ignimbrites quarried and used for construction.

Many of the quarries in ancient Rome were underground, partly to preserve the ground surface for other uses and partly for ease of cutting and removing blocks from below. Quarries are scattered throughout the city, with a great deal of tuff used not only in construction of buildings, but for defensive walls. The first wall around the city, mostly large blocks of tuff, was built in the sixth century B.C. by King Servius. Quarrying of a slightly stronger tuff resulted in quarry sites being located well beyond the limits of the city.

Modern quarrying of tuffs is accomplished by using gas-motor-driven hydraulic machines. The quarrying machine drives two rotary saws (vertical and horizontal) and is mounted on rails on the quarry floor. Cut blocks are kicked to the side, where they are stacked and loaded for transport. Modern quarries produce mostly smaller tuff blocks, which can be easily carried and placed by the stone mason. The size depends to some degree on the bulk density of the stone.

Quarry dust should be analyzed to determine if there are health hazards associated with the mineralogy of the tuff. The main problem is with silica polymorphs, including fine-grained quartz, tridymite, and cristobalite, which may be found in the quarry dust. The other minerals that can be carcinogenic are the fibrous zeolites, which can be found as secondary minerals in some tuffs.

Modern quarrying leaves a relatively smooth, flat quarry floor, which allows the quarry to be used for some other purpose once it has been abandoned.

UTILIZATION OF ABANDONED QUARRIES

In cities, such as Rome and Naples, where tuffs have been quarried for thousands of years, there are severe problems associated with the quarries in suburban areas. The lack of any kind of policy or systematic strategy in planning the utilization of abandoned quarries has led to misuse of valuable land and creation of geotechnical hazards or environmental problems associated with uncontrolled dumping. Civil governments for cities built on tuffs need to develop an approach to assess buried quarries in urban areas and, when identified, determine their stability and best use.

Abandoned quarries have several potential uses in urban and suburban areas, including: (1) Agriculture—for example, the value of agriculture on the Campanian Plain has led to the filling of quarries and the return of this reclaimed land to agriculture. Abandoned tuff quarries in Java have been reclaimed and are being used as rice paddies. (2) Resort areas—the quarries can supply a stable foundation for a hotel and garden or, with flooding, an area for water sports. (3) Sports facilities—quarry floors are ideal surfaces for tennis courts or football fields. (4) Water supply—lined quarries can be used as storage reservoirs. (5) Solar energy sites—quarries offer an unencumbered view of the sun and have stable surfaces for solar collectors and the control systems.

The main problems associated with the use of abandoned quarries include: (1) Quarry stability—a complete geotechnical analysis of the stability of quarry walls is required before using the site. (2) Water table intersecting quarry floors—this is a common problem for all quarries and not unique to those in tuffs. (3) Pollution in quarries and illegal dumping—abandoned quarries attract illegal dumping. Usually this is by local people who don't want to pay the fee for the local solid-waste landfill, but increasingly these quarries are being used by criminal organizations to dump toxic chemical wastes. Geologists who have become accustomed to visiting their favorite outcrops to study the history of an explosive volcano, find themselves driven out of the quarry sites by foul fumes and pools of unidentified chemicals. (4) Modification of countryside—quarrying radically changes the appearance of the landscape, and after closure, must be considered for another use or filled and soils restored.

Modern abandoned quarries in tuffs are regulated in Italy by a law that gives permission to quarry only with a guarantee of full rehabilitation to the previous environment (!). Regardless of the law, these abandoned quarries are still a difficult problem. This problem exists in not only classical volcanic areas (Naples, Rome, etc.), but also in some provinces where the countryside's character is an important factor (e.g., Colli Eugenei e Lessini and the volcanoes of northeastern Italy). Unfortunately, the lack of systematic survey is resulting in much illegal waste disposal activity. Around the main urban areas, quarries are filled with all kinds of waste. A specific assessment of each area with quarries is preferred by the Italian Geological Survey.

The ancient style of quarrying activity was concentrated close to urban areas and consisted of mainly extended tunnel systems, which produced volcanic materials for buildings (pozzolan and tuff). In parts of Naples and Rome, there are areas collapsing over ancient quarries in the heart of historical areas. Some of these tunnel systems are filled with unhealthy sewage waters, slowly flowing below highly populated sectors of the cities, whereas others are undiscovered.

On the borders of urban areas, tuff tunnels are extensively used for growing mushrooms. This activity, which has some economic value, is mostly in welded tuffs or pozzolan units of the Alban Hills, south of Rome. Abandoned welded tuff quarries are used for fishing where the quarry floors are below a water table.

Underground quarrying was common in historical Italian cities located on tuffs. The purpose of the quarries were: (1) catacombs and subsurface temples (Christian period and Roman Empire), (2) mining of construction material (from Etruscan-Roman to Medieval times), and (3) service tunnels for sewage and water.

All of these tunnels are producing surface instabilities in urban areas such as Rome and Naples. The larger subsurface cavities are 2–3 m high, with pillars left from previous quarrying activity. When in lithoid tuffs, the excavated chambers can be over 10 m high. The largest chambers are below Naples, where the tunnel system is deeper than in Rome. A subsurface survey of excavated chambers is absolutely necessary, not only downtown, but in the suburban areas where historical activity was extensive.

THE USE OF UNDERGROUND SPACE IN TUFF DEPOSITS

Sustainable development of cities will be aided by use of underground space. Limited land, constant growth, and high real estate values will eventually necessitate the use of this space in ways beyond subways and parking garages. The even temperatures of underground space, lower energy costs for heating and cooling, and such space can be used for offices, warehouses, and factories. Underground space centers have been established in Minneapolis, USA, Sydney, Australia, and Tokyo, Japan, where the focus is on the use of this resource in future cities. Cities located on especially thick tuff deposits can benefit from this resource, which is easily quarried and structurally stable.

Throughout the tuff plateaus of Turkey, Armenia, and Georgia, whole cities were excavated in thick, nonwelded or poorly welded ignimbrites because of the ease of excavation, excellent insulating properties, and potential for developing well-fortified locations.

In Japan, where the older "Green Tuff" deposits are located, there have been large-volume excavations for factories and storage areas.

There has been little published in the modern literature on use of underground facilities in tuff deposits. Yilmazer (1995) described the geological engineering required for an International Underground Congress Hall in Cappadocia, Turkey, near the town of Avanos. The hall will be excavated in a 9-m-thick ignimbrite, which is part of a sequence of tuffs, including nonwelded fallout units. The ignimbrite is much stronger than any of the underlying or overlying thin-bedded tuffs. The hall will be located in a "remnant" hill that is between older quarries. Near-vertical joints are widely spaced. This hall will be one of the largest underground structures in the world, with an area of 3200 m².

The future of technology for development of underground facilities in tuffs lies with the 8-km-long tunnel bored as part of an Exploratory Studies Facility to characterize rocks for the proposed high-level nuclear waste storage at Yucca Mountain, Nevada (N. Elkins, presentation at this workshop). The site is in a complex sequence of tuffs from a large Tertiary caldera complex and includes mostly ignimbrites with all degrees of welding, but also interbedded pumice and ash fallout deposits that have been cemented with zeolites and clays. The preliminary core hole drilling plus the detailed geological and engineering work done in association with this tunnel will provide the most detailed data set on tuffs available anywhere.

The main tunnel was started in 1994 by the excavation of a 10×9 m slot in the face of a small hill. This slot, excavated by conventional drill and blast methods, formed a launch chamber, 60 m in length. Into this launch chamber was inserted a 7.6 m tunnel boring machine (TBM).

The TBM and all of its attached trailing equipment weighed 860 tons and was 140 m long, including 13 work platforms, which were towed behind the TBM. The TBM operated by gripping the sides of the tunnel with its gripper shields and pushing forward its rotating mechanical cutter head, which had 48 cutting discs, to excavate the tunnel by chipping and grinding the tuff. The TBM was laser guided and electrically powered and had an operating voltage of 12.47 kV, providing 3800 horsepower and one million pounds of forward thrust. The maximum advance rate was 5.3 m/h. Concrete inverts were placed along the entire 8 km length of the tunnel floor to provide a flat surface for laying track and conducting all operations. The tunnel's crest was lined with steel mesh, which was held in place by rock bolts. The roof and walls were supported, depending on the specific ground conditions, by rock bolts, steel mesh, steel sheets, and, when needed, steel lagging bolted to the large girders of the steel sheets. The main tunnel was completed in 1997.

Bearing roughly east-west off of the north ramp of the horseshoe-shaped main tunnel, across the west side of the repository block, a cross-drift, 3 km long, was excavated with a 5-m-diameter TBM. In addition, eight alcoves and a number of smaller niches that were designed for scientific or engineering tests were excavated by mechanical miners (road headers or alpine miners) or by conventional drill and blast methods (Levich et al., 2000).

FOUNDATIONS IN TUFF DEPOSITS

Cities located on thick tuff deposits tend to have favorable seismic properties in terms of building stability. Morphological and geotechnical properties typical of tuff cities produce strong differences between seismic impedance of lower alluvial deposits and higher tuff cliffs or hills. Systematic data collected in Rome for 150 historical buildings or structures and macroseismic information from Naples demonstrate that there is severe damage to buildings on alluvial fill and along edges of tuff cliffs. Numerical models show the strong effect of subsurface geology on the amplification of seismic shaking in alluvium, in contrast to tuffs.

OTHER PYROCLASTIC CONSTRUCTION MATERIALS

Pumice and Ash

One of the oldest known construction materials, pumice, was used along with pozzolan (fine-grained or zeolitized ash) by the Romans as lightweight aggregate in concrete. Pumice continues to serve as a source of lightweight aggregate for construction, either in poured concrete or in pumice or cinder blocks. When mixed with cement, the lightweight aggregates produce blocks that have low density, good crushing strength, are fire resistant, moisture resistant, and are excellent acoustical insulators.

Pumice concrete is appropriate in construction where building weight must be minimized. If pumice sources are reasonably close, the savings in cost of structural steel justify the use of pumice concrete. Within poured concrete, pumice aggregates are uniformly distributed because they do not sink before the concrete hardens. Because of its elasticity, pumice concrete resists shock; this was demonstrated in Germany during WWII, where houses built with pumice concrete resisted bombings much better than those made with conventional concrete (Schmidt, 1956). Pumice aggregates also increase the insulating value of either poured concrete or masonry. For example, the "K" factor for concrete made with sand and gravel aggregate is 12.5, while that for pumice aggregate is 2.4. (K factors are btu/ft^2/hr/inch thickness/°F; the lower the K factor, the better the insulating properties; Schmidt, 1956.)

Fine pumice aggregates (<2.5 mm and >0.6 mm) are also used in plaster, which weighs a third less than conventional plaster. The weight savings for standard gypsum hard-wall plaster is 15 kg/m^3 (Schmidt, 1956).

The most common and easily mined pumice deposits are fallout deposits near their source vents. Most quarries are located on the slopes of tuff deposits that surround rhyolitic or dacitic domes (tuff rings) or in fallout deposits associated with caldera-forming eruptions (Plinian deposits). For example, thick pumice-fall deposits are quarried on the slopes of Glass Mountain, in northeastern California, where thicknesses range from 2 to 4 m (Chesterman, 1956; Heiken, 1978). Wall and slope stability is high, allowing open-pit quarrying with little or no structure required to maintain a vertical face. Pumice-fall deposits are easily quarried using front-end loaders and bulldozers. Pumice products are naturally well-sorted by size, with median grain sizes of 1–10 mm; sizing for aggregates requires little other than coarse screening to eliminate coarser lapilli, bombs, and blocks. Pumice fragments (pyroclasts) preferred for concrete aggregate should be equant to slightly elongate and have vesicularities between ~25% and 50%. Pumices containing few or no minerals (phenocrysts) are preferred, minimizing pyroclast densities. The pumice-fall deposits should also have low contents of rock fragments (lithic clasts).

Pozzolan

Pozzolans are natural volcanic silicates that react with lime in water to produce a strong concrete and are especially useful for constructing marine structures, such as breakwaters and piers. Pozzolan hydraulic cement was used by the Romans for the construction of public buildings such as the Pantheon in Rome, roads, and aqueducts; the name "pozzolan" comes from the town of Pozzuoli on the Bay of Naples (Sersale, 1958).

The main components for pozzolan cement are either fine-grained silicic volcanic ash or crushed zeolitized tuffs (pumice and ash cemented with zeolite and clay minerals). Pozzolans must be fine-grained (with a lot of reactive surface area), thus making most hydrovolcanic tuffs excellent resources, whether or not they are glassy or zeolitized. In countries without access to hydrovolcanic tuffs, pulverized fly-ash is often used for pozzolanic cement.

The Roman engineer Vitruvius specified a ratio of 1 part lime to 3 parts of pozzolan for cement used in buildings and 1 part lime to 2 parts pozzolan for underwater structures. The ratio for modern structures using pozzolan cement is more or less the same. The reaction between the ash or zeolitized tuffs used for pozzolan and lime is slow, but the cement becomes progressively stronger and more durable with time (Mielenz, 1950). Pozzolan cements are also resistant to chloride-ion penetration.

Hydrovolcanic tuffs (volcanic ashes formed by rapid quenching of magma by water) and cemented zeolites from the Phlegrean Fields, near Naples, were the main pozzolan resources for the Romans, although similar resources were available in and near Rome. Fine-grained hydrovolcanic tuffs that erupted during the seventeenth century B.C. caldera-forming eruption on the island of Thera, Greece (Santorini), were quarried as pozzolan during the mid-nineteenth century A.D. and shipped to Egypt for cement lining in the Suez Canal. Quarrying of these tuffs for pozzolan cement and lightweight concrete continued until recently, when quarrying ceased because of the greater value of the island for tourism. Vitric hydrovolcanic ashes can be used as pozzolan after disaggregation. Zeolitized tuffs must be crushed then heated to 700–800 °C before being used in pozzolan cement.

Natural pozzolan resources are not limited to fine-grained silicic or trachytic vitric tuffs. A zeolitized ignimbrite (the main phase being clinoptilolite) was mined near Tehachapi, California, and used for pozzolanic cement in construction of the Los Angeles aqueduct (Mielenz, 1950). Kitsopoulos and Dunham (1996) and Fragoulis et al. (1997) have proposed quarrying clinoptilolite-, heulandite-, and mordenite-bearing tuff deposits on the islands of Polyegos and Kimolos, Aegean Sea, Greece, for industrial use in pozzolan cement.

Mining tuffs for use in pozzolan cement requires little more than a bulldozer with a ripper and front-end loaders for quarrying vitric tuffs plus crushers for zeolitized tuffs. When the quarries on Thera, Greece, were active, ships would anchor next to the caldera walls in the flooded caldera; front-end loaders carried fine hydrovolcanic ash from the quarry face to the cliff edge, dumped the ash through a screen to separate the lithic clasts, and the ash would go down a chute directly into the ship. Quarrying wasn't always that simple; over much of the past 100 yr, quarrying of the thick deposits was accomplished by tunneling under the deposit

until it collapsed into the quarry; then the loose ash was carried to the cliff edge to be loaded into ships. Collapsing quarry faces and constant fine ash dust made pozzolan quarrying on Thera a hazardous profession.

Scoria

Strombolian and Vulcanian eruptions form scoria (cinder) cones of all sizes. The cones, composed of interbedded ash, lapilli (volcanic equivalent in size to pebbles), and bombs (equivalent in size to cobbles and boulders) of basaltic or basaltic andesite composition, are a resource of considerable value, especially if the cones are located near a city or railroad. The popularity of scoria as an aggregate is such that, for example, many of Auckland, New Zealand's, scoria cones have disappeared; only the inclusion of cones into city parks has preserved the volcanic heritage of this city.

Scoria is used mainly to build roads, both gravel and asphalt highways and railroad beds. Major quarries in scoria deposits near Flagstaff, Arizona, are there because of the coincidence of a railroad town with a volcanic field containing hundreds of scoria cones. Graded scoria is used mainly as a subbase and base for roads; 46% of the 970,000 t of scoria used in the state of Victoria, Australia, has been used for roadbeds (Guerin, 1992). However, when used as asphalt aggregate, there is excessive tire wear, and scoria should be used only for paved highways where high skid resistance is desirable. Scoria is not ideal for stock tracks, because it cuts the hooves of cattle (Guerin, 1992).

It is the skid resistance that makes very fine, ash-size scoria desirable for sanding icy roads. The sharp edges of vesicle walls in individual pyroclasts provide the needed traction on ice. After a major snow storm in northern New Mexico, roads are red from the application of oxidized ash-size scoria from quarries on scoria cones located outside of the city of Santa Fe, New Mexico, USA. Careful crushing and sieving is needed, for lapilli-size particles will be kicked up by tires and result in cracked or pitted auto windows.

Scoria cones are formed by alternating Strombolian and Vulcanian activity; a process of ballistic deposition and subsequent slumping or avalanching provides the sloping beds of naturally sorted scoria (McGetchin et al., 1974). These processes give the cones their value; the bouncing and sliding of scoria pyroclasts down cone slopes prevent most welding and actually cause some sorting of the scoria. The most complete study of a scoria cone is that of the Rothenberg cone, east Eifel, Germany (Houghton and Schmincke, 1989), where the outer slopes have the most valuable scoria resource in the form of avalanche deposits and ash fall. The vent area consists of solid dikes and welded and partly welded scoria and is not easily quarried.

Quarries or "scoria pits" are usually located on the lower slopes of cones, in the avalanche deposits, where fewer large bombs and blocks need to be screened out. Much of the scoria is not welded, and the headwall of the quarry can be pushed toward the center of the cone and stopped when material is too welded to be easily removed. Most of the quarrying can be done with a bulldozer, front-end loader, screens, and dump trucks. In many cases, "pit run" scoria can be used for roadbed construction with no screening required.

If the scoria cone is located near a town, there will be concern about the visual effect of a quarry. Guerin (1992) recommends that the quarries be kept near the base of a cone, preferably on the back side (away from the town) or hidden in irregular topography.

Scoria is occasionally used as aggregate in concrete blocks, when it is the only aggregate available. Large bombs and scoria are used as decorative rocks in gardens, especially those that are a bright red. Scoria is sold as a base in barbecues, with the dual purpose of retaining heat and absorbing grease from the meat being grilled.

ARTISTIC USES OF TUFFS (STATUES, ARCHITECTURAL MONUMENTS)

Tuffs have been used throughout history for artistic purposes, ranging from the monuments on Easter Island to trim on colonial churches in Latin America. However, contemporary uses of tuffs for stone carving are evident throughout the world wherever there are tuffs. Within the United States, garden shops and architectural supply houses everywhere have a selection of art objects, from stone columns to bird baths, all carved from tuffs in quarries located over much of western and central Mexico. The variety of tuffs are selected on the basis of color and texture and generally represent different facies, degrees of welding and postdepositional alteration of ignimbrite units. Tuffs provide the material for much of the stone-worker's products that are exported from Mexico and many other Latin American countries.

PRESERVATION OF TUFF BUILDINGS AND MONUMENTS

Early work by geologist Z.A. Atzagortzian on the longevity of volcanic tuffs in Armenia was an effort to encourage the use of tuffs as building stones. Atzagortzian's approach was to study tuff stability of 1000- to 1500-yr-old architectural monuments. He found that monuments constructed of moderately welded, moderately porous tuff had a patina that protected the stone, ensuring long life. Monuments constructed of denser welded tuff suffered from wetting, drying, and frost action. Later work by Atzagortsian and O.S. Martirosian was covered in the monograph *Tuffs and Marbles of Armenia* (1962). Atzagortzian and Martirosian concluded that restorers should leave the monuments alone, use no coatings of any sort, and use the natural patina as the preservative.

The deterioration and possible treatments of tuff stone work have been summarized by Grissom (1994). Grissom discusses famous historical and archaeological sites, such as Easter Island (colossal heads carved from basaltic hydrovolcanic tuff), Borobudur, Indonesia (33,100 relief blocks and 500 sculptures carved using fine-grained, nonwelded ignimbrite), and churches carved from ignimbrites at Göreme, Turkey, and Lalibela, Ethiopia. On Easter Island, deterioration of the monolithic heads is mostly caused by rainfall; the growth of lichens and algae also speeds the weathering process. The general opinion among preservationists is that these monuments should be treated with water repellents and fungicides to slow their deterioration.

Tuffs have been used throughout history for construction in central Italy, from the Etruscans to the present (see Chapter 6 by de Rita et al., this volume). The large variety of tuff compositions and degrees of welding and cementation make it nearly impossible to establish a uniform treatment to stabilize monument surfaces. J. Griffin found that tuff stabilization in Italy involves mostly a combination of a biocide and stabilization-waterproofing compounds (used for ancient Roman and fifteenth century palaces). Most experience with treatment of tuffs has been in the laboratory and not in the field. Air pollution seems to have little effect on tuff monuments, and the main disfiguration of carved tuff is by biological growth (Grissom, 1994).

REFERENCES CITED

Atzagortsian, Z.A., and Martirosian, O.S., eds., 1962, Tufy i mramory Armeni (Tuffs and marbles of Armenia—in Russian): Erevan, Armenian Institute of Materials, 157 p.

Chesterman, C.W., 1956, Pumice, pumicite [volcanic ash], and volcanic cinders in California: California Division of Mines and Geology Notes, v. 174, p. 3–98.

Fenner, C.N., 1948, Incandescent tuff flows in southern Peru: Geological Society of America Bulletin, v. 59, p. 879–893.

Fragoulis, D., Chaniotakis, E., and Stamatakis, M.G., 1997, Zeolitic tuffs of Kimolos Island, Aegean Sea, Greece, and their industrial potential: Cement and Concrete Research, v. 27, p. 889–905, doi: 10.1016/S0008-8846(97)00072-0.

Grissom, C.A., 1994, The deterioration and treatment of volcanic stone: A review of the literature, *in* Charola, A.E., Koestler, R.J., and Lomardi, G., eds., Lavas and volcanic tuffs: Rome, International Center for the Study of the Preservation of Cultural Property, p. 3–29.

Guerin, B., 1992, Review of scoria and tuff quarrying in Victoria: Victoria Geological Survey Report 96, p. 1–72.

Heiken, G., 1978, Plinian-type eruptions in the Medicine Lake Highland, California, and the nature of the underlying magma: Journal of Volcanology and Geothermal Research, v. 4, p. 375–402.

Holmes, W.H., 1985, Archeological studies among the ancient cities of Mexico: Field Colombian Museum, Publication 8, Anthropology Series, v. f1, no. 1, 338 p.

Houghton, B.F., and Schmincke, H.-U., 1989, Rothenberg scoria cone, East Eifel: A complex Strombolian and phreatomagmatic volcano: Bulletin of Volcanology, v. 52, p. 28–48, doi: 10.1007/BF00641385.

Jenks, W.F., and Goldich, S.S., 1956, Rhyolitic tuff flows in southern Peru: The Journal of Geology, v. 64, p. 156–172.

Kitsopoulos, K.P., and Dunham, A.C., 1996, Heulandite and mordenite-rich tuffs from Greece: A potential source for pozzolanic materials: Mineralium Deposita, v. 31, p. 576–583, doi: 10.1007/s001260050064.

Levich, R.A., Linden, R.M., Patterson, R.L., and Stuckless, J.S., 2000, Hydrologic and geologic characteristics of the Yucca Mountain site relevant to the performance of a potential repository, *in* Lageson, D.R., Peters, S.G., and Lahren, M.M., eds., Great Basin and Sierra Nevada: Boulder, Colorado, Geological Society of America Field Guide 2, p. 383–414.

McGetchin, T.R., Settle, M., and Chouet, B.H., 1974, Cinder cone growth modeled after Northeast Crater, Mt. Etna, Sicily: Journal of Geophysical Research, v. 74, p. 3257–3272.

Mielenz, R.C., 1950, Materials for pozzolans: A report for the engineering geologist: U.S. Bureau of Rec., Petrographic Laboratory Report Pet-90B, 25 p.

Nelson, P.H., and Anderson, L.A., 1992, Physical properties of ash flow tuff from Yucca Mountain, Nevada: Journal of Geophysical Research, v. 97, p. 6823–6841.

Funiciello et al.

Orsi, G., de Vita, S., and di Vito, M., 1996, The restless, resurgent Campi Flegrei nested caldera (Italy): Constraints on its evolution and configuration: Journal of Volcanology and Geothermal Research, v. 74, p. 179–214.

Schmidt, F.S., 1956, Technology of pumice, pumicite [volcanic ash], and volcanic cinders: California Division of Mines and Geology Notes, v. 174, p. 99–117.

Sersale, R., 1958, Genesi e constituzione del tufo giallo Napolitano: Rendiconti Accademia Scienze Fisiche Matematiche, Classe di Scienze Fisiche, Matematiche e Naturali, v. 25, p. 181–207.

Toshioka, T., Tsuchida, T., and Sasahara, K., 1995, Application of GPR to detecting and mapping cracks in rock slopes: Journal of Applied Geophysics, v. 33, p. 119–124, doi: 10.1016/0926-9851(94)00022-G.

Wohletz, K.H., 1996, Fracture characterization of the Bandelier Tuff in OU-1098 (TA-2 and TA-41): Los Alamos National Laboratory Report LA-131194-MS, 19 p.

Yilmazer, I., 1995, Engineering geology factors in the design of a large underground structure in a tuff sequence in Cappadocia: Engineering Geology, v. 40, p. 235–241, doi: 10.1016/0013-7952(95)00057-7.

MANUSCRIPT ACCEPTED BY THE SOCIETY 29 DECEMBER 2005

Geological Society of America
Special Paper 408
2006

Chapter 6

A case study—Ancient Rome was built with volcanic stone from the Roman land

Donatella de Rita
Ciriaco Giampaolo
Dipartimento di Scienze Geologiche, Universita degli Studi di Roma–TRE, Largo San Leonardo Murialdo 1, 00146 Roma, Italy

RECENT GEOLOGICAL EVOLUTION OF THE ROMAN REGION

The Roman territorial framework is mostly the product of recent geodynamic processes affecting the western Mediterranean area. Rome's foundations are mostly the products of the Quaternary explosive eruptions in two volcanic fields, the Sabatini volcanic field (northwest), and the Colli Albani (Alban Hills) volcanic district (southeast). Volcanoes in both volcanic fields emplaced sequences of pyroclastic flows, the deposits of which formed plateaus deeply excavated by postglacial erosional processes.

About 1 million years ago, the Roman land was still submerged under the Pliocene sea; outcrops of clay and sandy-clay sedimentary rocks from this period (3.40–1.79 Ma) make up the highest part of the city, including the Vatican, Janiculum, and Monte Mario hills. These sediments were deposited in basins that developed as a consequence of the Apenninic orogeny and the subsequent opening of the Tyrrhenian Sea (Funiciello, 1995). They were uplifted to the present position by tectonic processes concurrent with volcanism. The Pliocene claystones, with a thickness of more than 800 m, constitute the bedrock of the Roman area; their tectonic and stratigraphic relationship with the younger volcanic sediments have strongly affected the urban development of Rome, and they are responsible for some of the problems affecting the land stability of entire sectors of the city.

The Roman area remained submerged until 0.88 Ma, but during this time interval, variations of sea level and tectonic processes created a setting for shallow-water sedimentation. Three main marine cycles have been identified during this time interval. The first, from 3.40 to 1.79 Ma, refers to the deposition of clay-marly sediments ("Marne del Vaticano;" Marra et al., 1995); the second to sandy and sandy-clay sediments ("Unità di Monte Mario;" Marra, 1993); and the third to the deposition of infra-littoral clays ("Unità di Monte delle Piche;" Marra, 1994). The second and the third marine cycles occurred between 1.79 and 0.88 Ma and are separated by continental sediments that indicate a period of emergence (Funiciello, 1995).

At ca. 0.88 Ma, continental sedimentary conditions were dominant in the Roman area (Funiciello et al., 1992). This was an important step in the recent evolution of the Roman land, because from that time, the area was shaped by erosional and depositional processes in a manner similar to that at present. As an example, the course of the Tiber River was established. The previous Tiber River was east of its present course and was shifted by the effects of the extensional tectonism, which affected the coastal area of Latium, by glacio-eustatic variations of sea level, and by the emplacement of the oldest pyroclastic flows erupted from the Sabatini and the Colli Albani volcanic fields. The volcanic activity of these two districts began ca. 0.6 Ma and emplaced several sequences of pyroclastic flows, most of which reached the Roman area and formed large flat, plateaus.

Finally, fluvial erosion processes during the Würm glacial period eroded deep valleys into the plateaus. This is the origin of the famous seven hills of Rome.

The oldest part of the city was at the foot of two of these hills, the Campidoglio (Capitoline) and the Palatino (Palatine), next to the Isola Tiberina, in a place with easy access to the Tiber for commercial markets. Eventually the Romans moved to the hilltops for security and better climatic conditions. The geological setting was the determining factor for rapid development of Rome and the influence that made Rome one of the most famous cultures of the world.

de Rita, D., and Giampaolo, C., 2006, A case study—Ancient Rome was built with volcanic stone from the Roman land, *in* Heiken, G., ed., Tuffs—Their properties, uses, hydrology, and resources: Geological Society of America Special Paper 408, p. 127–131, doi: 10.1130/2006.2408(06). For permission to copy, contact editing@geosociety.org. ©2006 Geological Society of America. All rights reserved.

VOLCANIC STONE USED FOR BUILDING FROM THE SABATINI AND COLLI ALBANI VOLCANIC FIELDS

Much of Rome's geologic foundation consists of tuffs (ignimbrites) from the Colli Albani volcanic field, with lesser volumes of ignimbrites from the Sabatini volcanic field in the northwestern part of the present city. The first tuffs in the Roman area were fall deposits, partially reworked, from the Sacrofano volcano (Sabatini volcanic district). They crop out in the northwestern edge of Rome. The first thick tuff deposit useful in construction is the "tufo pisolitico," which is the oldest pyroclastic flow deposit erupted from the central volcano (Tuscolano-Artemisio) of the Colli Albani district (de Rita et al., 1988). The tufo pisolitico was not deposited as a single unit but as a sequence of pyroclastic flows erupted during different cycles of the Tuscolano-Artemisio caldera complex (0.6 Ma) (de Rita et al., 1988, 1995, 2002; Rosa, 1995). These eruption cycles were separated by long periods of quiescence, as evidenced by the presence of paleosols. During the eruptions, there were magma-water interactions as evidenced by: (1) the fine grain size of the units, (2) the presence of accretionary lapilli (hence name of "pisolitico"), (3) the presence of trunks of trees or of their marks, (4) cross-bedded stratification of the pumice layers, and (5) zeolitic alteration of the glassy matrix (chabazite and phillipsite; Fornaseri et al., 1963). The eruptions produced moderate volumes of pyroclastic flows, which flowed as far as the Tiber valley with resulting thicknesses of 5–10 m (de Rita et al., 1995)

Concurrent with the Tuscolano-Artemisio activity, very similar pyroclastic flows were erupted in the Sabatini volcanic district and reached the northwestern part of what is now Rome. These are the yellow "Via Tiberina tuff," which is a pyroclastic flow deposit from the Sacrofano volcano, located 30 km north of Rome, in the eastern sector of the Sabatini volcanic field (de Rita et al., 1993). The Sacrofano eruption, dated at ca. 500,000 yr B.P., likely involved some interaction of water with rising magma. The eruption produced at least seven units, which covered a surface of more than 400 km^2 and had a total volume of ~8 km^3 (Cioni, 1993; de Rita et al., 1993; Rosa, 1995). The Sacrofano pyroclastic flows caused a significant impact on the surrounding environment, such as the obstruction of the Tevere (Tiber) River near Monte Soratte, where the valley was shifted to the east, approximately coincident with the present course of the river (Alvarez, 1972, 1973; Rosa, 1995). After the emplacement of these units, there was a period of quiescence during which the Roman area was affected by tectonism and intense local fluid circulation depositing travertine and altered sands. Probably this fluid circulation and alteration was facilitated by the presence of N-S faults that were partly responsible for the deviation of the river's course (Funiciello, 1995).

After that, with a cycles of ~100 k.y. (de Rita et al., 1993), a sequence of pyroclastic flows from the Colli Albani volcanic district reached Rome. The oldest of these units is called "pozzolane rosse" or "Il colata piroclastica del Tuscolano-Artemisio" and represents the most important explosive unit during the history of the volcanic field, with an eruption volume of more than 34 km^3 (de Rita et al., 1988). The deposit consists of reddish-pinkish scorias in a loose scoriaceous matrix with free crystals of leucite, pyroxene, and biotite and thermo-metamorphosed sedimentary lithic clasts. Pipe structures (fossil fumaroles) are common, even far from the vent localities, e.g., those inside the present city area (more than 20 km from the source area). This unit, together with the younger "pozzolane nere," which is lithologically very similar, has been quarried since early Roman time and used to make concrete. The "pozzolane rosse" and the "pozzolane nere" deposits are separated by reworked tuffs and alluvial sediments, indicating a quiescent period between the two eruptions.

Finally, at ca. 0.4 Ma, a complex eruption in the Colli Albani deposited two more important flow units, the "Lionato tuff" and the "Villa Senni tuff" (336,000 yr B.P.), which concluded the activity of the Tuscolano-Artemisio volcano, causing the collapse of the central part of the edifice (de Rita et al., 1988; Funiciello et al., 1995). Both of these pyroclastic flow units reached the area of Rome. The lithological and depositional characteristics of the "tufo lionato" indicate that there was sustained but limited water-magma interaction during eruption, and that its high degree of lithification was caused by zeolitic alteration of the glassy matrix (herschelite, chabazite, and phillipsite; Fornaseri et al., 1963).

After 336,000 yr B.P., the area occupied by Rome was not again covered by pyroclastic deposits because the volcanic activity of the Colli Albani and of the Sabatini volcanic districts began to wane and smaller-scale explosive activity affected only the areas of the volcanic fields. Southeastern Rome, close to the tomb of Cecilia Metella, was reached by a large lava flow ("Colata di Capo di Bove"). This lava flow erupted ~280,000 yr B.P. from the Le Faete edifice in the central area of the Colli Albani volcanic district (Bernardi et al., 1982; de Rita et al., 1988). The lava was channeled in a valley nearly radial from the central volcanic edifice and flowed 20 km to the edge of the area on which Rome was eventually built. The flow was named after the ox head perched on the tomb of Cecilia Metella, which was built at the distal end of the flow. Cecilite was the previous scientific name (after Cordier, 1868) of this aphyric variety of melilite leucitite. The highly undersaturated chemistry of the lava permitted it to flow a great distance and to form a very smooth and flat upper surface, which was directly exploited by the Romans for the foundation of the ancient Appian Way road.

The deposits described herein were used by the Romans for construction materials; loose pozzolan was used in the preparation of cement, and the lithified nonwelded ignimbrite was used as building blocks. But Romans, during the first part of the Imperial period, also used building blocks that were quarried far from the city. The most common tuffs used for building blocks are famous, called "peperino," which refers to two different lithotypes: the lapis Gabinus and the lapis Albanus. Both of these deposits are the result of violent hydromagmatic eruptions, the former related to the Gabii or Castiglione crater and the later related to the Albano crater. The Castiglione and Albano craters, on the slopes of the

Colli Albani, were formed during the most recent hydromagmatic activity of the Colli Albani volcanic district (de Rita et al., 1988, 1995). This region was still active ca. 200,000 yr B.P. and also very recently: in particular lapis Albanus came from a detrital flow deposit erupted from the Albano crater during its final phase, at least 20,000 yr ago (Mercier, 1993). The two deposits have very similar lithologies, and, as their principal characteristic, a high level of lithification because of zeolites in the rock matrix derived from the alteration of volcanic glass. Both rocks have chabazite, phillipsite, and harmotome in the zeolitic matrix. These rocks were used everywhere in Rome during the Imperial epoch. Late in the Imperial period, another stone, named sperone, was used for the construction of the most representative monument of the epoch: the Coliseum. Sperone is a welded scoria deposit formed by lava fountains erupted from the fractures that controlled the collapse of the central part of the Colli Albani volcano less than 336,000 yr ago (de Rita et al., 1988, 1995). These rocks compose the entire northern border of the Tuscolano-Artemisio belt and are located at an elevation between 200 and 600 m.

THE USE OF THE BUILDING STONES IN THE DEVELOPMENT OF ROME

An ancient morphological map clearly shows that Rome developed mainly on the southeastern margin of the present city and that the seven hills were all located on the east side of the Tiber valley. This part of Roman territory was underlain by volcanic deposits from the Colli Albani district, which overlay clay sediments of Pliocene-Pleistocene age. The seven hills were isolated by deeply eroded valleys occupied by river courses flowing directly into the Tiber. The sedimentary substratum is exposed in deeply eroded valleys; many of these outcrops are obscured because of urbanization. During the Archaic period, common erosional processes affecting the sides of the valleys of the seven hills produced large erosion blocks that could be used for many purposes, including building stones, defensive walls, or shelters. It is then not surprising that the oldest structures of the city were constructed of tuff blocks excavated in areas within the city walls.

The Tufo Pisolitico or Cappellaccio

During the Archaic period the most-used stone was the cappellaccio, or tufo pisolitico, from the oldest pyroclastic flows erupted from Tuscolano-Artemisio volcano of the Colli Albani district. The name "cappellaccio" has been improperly used to indicate different tuffs; quarrymen used this name for any friable material, which was in some places weathered to soil without specific reference to a particular rock type or stratigraphic unit. This is probably the reason for the confusion, especially in the archaeological literature, about this name and the stratigraphic units. In some cases, the same name refers to building stones of the litoide, or lionato pyroclastic flow unit, which is another deposit that erupted from the central area of the Colli

Albani volcanic field in a later eruptive cycle and is present in some monuments of later periods. For these reasons, we prefer the name of "tufo pisolitico," and we propose to drop the "cappellaccio" term. During the Archaic period, Romans used mainly blocks from this tuff for the infrastructure of the city. Although the physical-mechanical characteristics of the pisolitic tuff are definitely inferior to those of subsequently used building stones, the ease of excavation and limited amount of transport made this material an efficient choice during the Archaic period.

The tuff units are thickest in the paleovalleys, and, in some cases, totally filled the valleys (de Rita et al., 1992). Tuff-filled paleovalleys are still visible in outcrops near the historical center of Rome. For example, in the area of the Campidoglio (Capitoline Hill), outcrops of the tufo pisolitico and the upper litoide allow reconstruction of the recent geological history of this area. Here, the relationships between the outcrops suggest that the pisolitico filled a paleovalley located approximately in the ruins of the Foro di Cesare (Forum of the Caesars). The valley was totally filled with tuff deposits, and the river course was shifted toward the present Via dei Fori Imperiali. After a long period (from 0.5 to 0.3 Ma), when no other pyroclastic flows reached the Campidoglio (Capitoline Hill), the litoide pyroclastic flow was deposited, filling the new valley. The pisolitico and the upper litoide ignimbrites were isolated and are now visible in the cliffs that surround the Capitoline Hill.

Many examples of the tufo pisolitico blocks are still visible in Rome, but probably the best example is the Mura Serviane (Servian Walls), which was the first defensive wall around Rome. Some of the quarries that supplied the building stone for the walls are still recognizable in archaeological excavations below Termini railway station.

Many other works of the Archaic period, preserved near the Foro di Cesare (Forum of the Caesars), were constructed with this stone; for example, an Archaic cistern was excavated in the tuff, and its walls were lined with small blocks of the same material (opus quadratum; Lugli 1957). In addition, the earliest hut village at the top of the Palatino (Palatine) hill was constructed from pisolitic tuff.

The Tufo Giallo Della Via Tiberina

With improving technological ability and an expanding territory, small blocks of tuff from outside the city walls began to be used in Rome. One of the first rock types that was substituted for the pisolitic tuff was the yellow Via Tiberina tuff, which crops out north of Rome at Grotta Oscura (the Grotta Oscura tuff; Coarelli, 1974, and bibliography therein) along the Via Tiberina.

Because the physical-mechanical characteristics of the Via Tiberina tuff are significantly better than those of the pisolitic tuff (Nappi et al., 1979), the Via Tiberina tuff was commonly used as an ornamental stone or as a building stone for houses. It is important to note that the use of this tuff became common only after the Roman conquest of Veio, the Etruscan city that dominated the region rich in this natural resource (Coarelli, 1974). The use of this tuff in Roman buildings during the entire Roman period is still well documented, as the monument called Ara di Cesare in the Caesar Forum, but the first use of this tuff was for the restoration of the Servian Walls. These walls were largely restored with this stone after 396 A.D., following the damage caused to the original wall by the Gaelic invasion. Where presently visible, the wall is constructed of rows of 59-cm-high blocks, alternatively placed horizontally and vertically, thereby creating a structure that is up to 10 m high and sometimes greater than 4 m thick (Coarelli, 1974, and bibliography therein). The restoration took place in various locations simultaneously, as testified by the reconstructed rock junctions that do not always fit together perfectly. The total length of the wall has been calculated at around 11 km, encompassing a surface of 426 hectares, and thus enclosing the largest city on the Italian peninsula.

The Capo di Bove Lava Flow (Melilite Leucitite)

The Roman ability to obtain building stones with optimal physical-mechanical characteristics was great during the Republican Period (end of the fourth and third centuries B.C.), as demonstrated by the extensive use of lava blocks for paving stone. The Via Appia, a consular road between Rome and the Colli Albani area, was built during this period directly on the upper surface of the Capo di Bove lava flow. The use of lava blocks for paving stone became common, and many examples of lava roads are still visible inside and outside of the city. Lava blocks are still used for paving city streets. At the present, these small lava blocks are called Sanpietrini and have the shape of a truncated square pyramid.

The Peperino and the Lionato Tuff

The most interesting period relative to the goals of this study is Imperial Rome, a time span that marked the beginning of a long period of prosperity for Roman civilization. During this time, Rome greatly extended its dominion across the entire Mediterranean area, and as a result, the Romans introduced many "exotic" stones for ornamentation and construction of both their public and private buildings. However, many of these buildings still used local building stones for their foundations and internal structures. The higher technological ability allowed a more rigorous selection of lithotypes. The monuments of this epoch used three principal types of volcanic building stone from the Roman area: the lionato (or litoide), the peperino, and the sperone tuffs.

The most significant examples of these tuff blocks are visible in the Imperial Forums. In the most ancient of these forums, that of Caesar, the Temple of Antonino and Faustina was constructed in part with lapis Albanus. In contrast, the Forum of Augustus was largely created using lapis Gabinus and the lionato tuff (or litoide), with the latter unit also being used for the forum brick work and as a base for the temple of Marte Ultore. In addition, lapis Gabinus forms the foundation and walls of the Tabularium

building at Campidoglio. The contemporaneous use of the lapis Gabinus and the lionato tuff suggests that the two rock types were from adjacent quarries in the volcanic field. In fact, the lapis Gabinus quarries were located along the border of the Castiglione crater (in the northern sector of the district near the Aniene River), not far from the lionato tuff quarries located along Via Tiburtina. It is highly probable that the Aniene River was the most direct route for transporting these two rock types from the countryside into the city. The lionato stone used during the Roman epoch was most likely excavated in the Settecamini area (tufo dell'Aniene); other quarries were located closer to the city at the foot of the Monteverde hill (tufo di Monteverde). The tuff quarried at this last locality was used to a limited extent, probably because of its low quality and also because the quarries were located in a dangerous locality. In fact, the hill of Monteverde, composed of the lionato tuff directly overlying clay and alluvial sediments, is continuously subject to slides. The Roman quarries of the lapis Albanus were located along the northern borders of the Albano crater, near the city of the same name, or in the valley below the town of Marino. The lapis Albanus differs from the lapis Gabinus because it was excavated until the end of the most recent epoch and is still quarried to a small extent today.

The Sperone Tuff

Finally, the monument that represents the most important symbol of Imperial Rome is the Coliseum. The structural base of the Coliseum was built with large blocks of travertine and sperone, a rock type that was used infrequently. Although no definitive Roman quarries have been found, it is highly probable that the original sources were located near the present towns of Grottaferrata and Frascati in the Colli Albani and close to the principal transportation routes to Rome. These quarries have been obliterated by the growth of these two cities. The densely welded scoria (sperone) has optimum physical and mechanical properties, and it is likely that its use was limited only because of the difficult access and transportation to and from the quarries. All the depositional and lithological characteristics of the sperone can be clearly observed in the Tuscolo area, one of more impressive localities related to the history of Rome in the Colli Albani volcanic field. The perfectly preserved remains of a Roman village and a small theater built entirely with the sperone are still visible at this location.

After the Imperial epoch, the study of volcanic rocks used as building stones in Rome becomes almost impossible, as the Romans began to use manufactured bricks as their most important construction materials. Furthermore, they commonly reused the stones of more ancient monuments that were in ruins, deteriorated with age, or demolished by subsequent emperors. The borders and power of the Roman Empire were by this point so vast that the importation and use of stone from around the world were common. Beyond this point, the link between man and his environment extended to the vast regions that reflected the power of Rome.

CONCLUDING REMARKS

The use of volcanic stone as a construction material served the needs through time of the evolving civilization, representing the inseparable bond between man and nature. In this chapter, we identify seven important volcanic building stones that were used for the construction of many of the important monuments in ancient Rome. The types of stones used by the Romans to build, protect, and beautify the city closely followed their society's technological development. With time, rocks with increasingly better physical-mechanical characteristics were chosen and excavated at increasingly longer distances from building sites. Each monument contains the history of a period or of a fundamental phase of the evolution of Roman civilization.

The rock types used for construction show comparable physical-mechanical characteristics (Penta, 1956); all have an elevated level of lithification because of zeolites in the rock matrix (minerals which formed during alteration of the glassy matrix). Furthermore, all units were produced by eruptions that involved the interaction of rising magma with groundwater. These observations become extremely interesting in the research and characterization of the building stones and underline the necessity of mankind to clearly understand his environment in order to obtain the best benefits.

REFERENCES CITED

Alvarez, W., 1972, The Treia valley north of Rome: Volcanic stratigraphy, topographic evolution and geological influences on human settlement: Geologica Romana, v. 11, p. 153–176.

Alvarez, W., 1973, Ancient course of the Tiber river near Rome: An introduction to the middle Pleistocene volcanic stratigraphy of central Italy: Geological Society of America Bulletin, v. 84, p. 749–758, doi: 10.1130/0016-7606(1973)84<749:ACOTTR>2.0.CO;2.

Bernardi, A., de Rita, D., Funiciello, R., Innocenti, F., and Villa, M., 1982, Geology and structural evolution of Alban Hills volcanic complex, Latium, Italy: Abstract of the Workshop on Explosive Volcanism, San Martino al Cimino, Italy: Consiglio Nazionale delle Ricerche, Italia and National Science Foundation, USA.

Cioni, R., 1993, Il complesso di Bolsena e il vulcanismo alcalino-potassico del lazio settentrionale [Tesi di Dottorato V ciclo]: Pisa, Università degli Studi di Pisa, 236 p.

Coarelli, F., 1974, Guida archeologica di Roma: Venezia, Arnoldo Mondadori Editore.

Cordier, P.L.A., 1868, Description des roches composant l'ecorce terrestre et des terrains cristallins: Ed. Ch. D'Orbigny Savy Paris, 553 p.

de Rita, D., Funiciello, R., and Parotto, M., 1988, Carta geologica del complesso vulcanico dei Colli Albani: Roma, Consiglio Nazionale delle Ricerche, scale 1:50,000.

de Rita, D., Funiciello, R., and Rosa, C., 1992, Volcanic activity and drainage network evolution of the Colli Albani area (Rome, Italy): Acta Vulcanologica, v. 2, p. 185–198.

de Rita, D., Funiciello, R., Corda, L., Sposato, A., and Rossi, U., 1993, Volcanic units, *in* Di Filippo, M., ed., Sabatini volcanic complex: Consiglio Nazionale delle Ricerche, Quaderni della Ricerca Scientifica, v. 11, p. 33–78.

de Rita, D., Faccenna, C., Funiciello, R., and Rosa, C., 1995, Stratigraphy and volcano-tectonics, *in* Trigila, R., ed., The volcano of the Alban Hills: Roma, Raffaello Trigila, p. 33–71.

de Rita, D., Giordano, G., Esposito, E., Fabbri, M., and Rodani, S., 2002, Large volume phreatomagmatic ignimbrites from the Colli Albani volcano (middle Pleistocene, Italy): Journal of Volcanology and Geothermal Research, v. 118, p. 77–98, doi: 10.1016/S0377-0273(02)00251-2.

Fornaseri, M., Scherillo, A., and Ventriglia, A., 1963, La regione vulcanica dei Colli Albani: Vulcano Laziale: Roma, Consiglio Nazionale delle Ricerche.

Funiciello, R., (coordinatore scientifico), 1995, La geologia di Roma, Il centro storico: Memorie Descrittive Carta Geologica d'Italia, v. L, 550 p.

Funiciello, R., Giuliani, R., Marra, F., Salvi, S., 1992, Superfici strutturali plioquaternarie al margine sud-orientale del Distretto Vulcanico sabatino: Studi geologici Camerti, volume speciale 2 (1991), Progetto Crosta Profonda (CROP) 11, p. 301–304.

Lugli, G., 1957, La tecnica edilizia Romena con particolare riguardo a Roma e Lazio: Roma, Bardi editore, v. I, p. 169–359.

Marra, F., 1994, Stratigrafia ed assetto geologico-strutturale dell'area romana compresa tra il Tevere e il Rio Galeria: Geologica Romana, v. 29 p. 515–535.

Marra, F., Rosa, C., De Rita, D., Funiciello, R., 1995, Stratigraphic and tectonic features of the Middle Pleistocene sedimentary and volcanic deposits in the Roman Area (Italy): Quaternary International, v. 47/48, p. 51–63.

Mercier, N., 1993, The thermoluminescence dating technique: Applications and possibilities [abs.]: Rome, Symposium "Quaternary stratigraphy in volcanic areas," September 20–22, 1993, p. 2.

Nappi, G., De Casa, G., and Volponi, E., 1979, Geologiae caratteristiche tecniche del "tufo giallo della Via Tiberina": Bollettino Società Geologica Italiana, v. 98, p. 431–445.

Penta, F., 1956, Materiali da costruzione del Lazio: La Ricerca Scientifica, v. 26, p. 11–201.

Rosa C., 1995, Evoluzione geologica Quaternaria delle aree vulcaniche laziali: Confronto tra il settore dei Monti Sabatini e quello dei Colli Albani [Tesi di Dottorato di ricerca in Scienze della Terra]: La Sapienza, Università degli studi di Roma La Sapienza, VII ciclo.

MANUSCRIPT ACCEPTED BY THE SOCIETY 29 DECEMBER 2005